文化传播·创意设计·展览系列

致敬中国建筑经典

中国20世纪建筑遗产的事件·作品·人物·思想

中国文物学会20世纪建筑遗产委员会 《中国建筑文化遗产》编辑部 主编

天津大学出版社
TIANJIN UNIVERSITY PRESS

写在前面

It is our consistent pursuit to tell the "stories" of architectural culture, urban art, creative design and heritage protection with exhibitions, which is necessary for not only reading architecture but also appreciating urban culture. Zhang Qinnan, a critic of Chinese architectural culture, repeatedly emphasizes that "Architecture is a mirror of the times" in many of his works such as *Reading Architecture* and *Reading City*, as architecture grows with people and "integrates the soul of its creator and user in the architecture". He also asks specially, "What is the silent building telling us?" I think that exhibition to both insiders and outsiders is a good way to spread architectural culture, so we managed to plan this series of works on the basis of past practices.

The most valuable exhibition recently was the "Tribute to the Chinese Architecture Classics – Events, Projects, Figures and Thinking of the 20th Century Chinese Architectural Heritage" held at the 9th China Weihai International Habitat Festival in Shandong Province from September 15th to 17th, 2017. It benefited from the first edition of the 20th Century Chinese Architectural Heritage List, released jointly by China Cultural Relics Academy and the Architectural Society of China on September 29, 2016. If history is everlastingly new by recording, and the works of architects are unfading in the hearts of people by creation, then the modern classical architecture that the 20th Century Chinese Architectural Heritage presents and the collective and personal history of the architects can be called the "iconography" of a country's architectural history. Co-planed by experts from home and abroad, the exhibition was themed with "Tribute to the Chinese Architecture Classics" and combined the elements of events, projects, figures and thinking following the main line of the 20th century. With as much as innovation (actually the traditional mode), the exhibition presented the concept and people of Chinese architectural heritage projects in the 20th century in the form of art, culture or history, and explored the literature method to record the whole change course of the Chinese architecture in the 20th century. The exhibition was well received by both insiders and outsiders of the architectural world.

用展览讲好建筑文化、城市艺术、创意设计遗产保护等方面的“故事”，是我们的一贯追求，这不仅是阅读建筑，更是品味城市文化的需要。中国建筑文化的评论大家张钦楠先生在其《阅读建筑》《阅读城市》等多部著作中，一再强调“建筑是时代的镜子”，因为建筑与人一同成长，所以“建筑中便熔铸了它的创作者与使用者的灵魂”。他还特别发问：“沉寂的建筑在向我们诉说什么？”我理解，面向业界内外的展览是一种传播建筑文化的好形式，于是我们在过去实践的基础上鼓足勇气策划了这套丛书。

国内外近年来涉及城市、建筑、艺术设计历史与文化的展览很多，我们提出的不是一般的博物馆、美术馆乃至与档案馆图书馆相联系的展陈，而是专门服务于城市文化复兴、城市事件、建筑文化论坛、历史重大设计问题追踪等所办的展览。要看到，与国外“井喷”般文化展览业态格局相对比，国内的展览市场资源还有待进一步挖掘，总体上讲，涉及建筑文化、城市艺术、设计创意的文化能量还未充分释放。国内某些博物馆、美术馆展览内容不变，展陈手段单一，不考虑观众的接受程度，粗浅的标识下观众无法了解展览背后的信息，难以体味到展览背后的故事，这是展览策展意识差与策展能力欠缺的表现。

德国艺术策展大家鲍里斯·格洛伊斯在《走向公众》一书中，通过“联合国广场”和“夜校”等艺术系列讲座及展会，讲述如何用普及方式将艺术观传递给观众。诚如设计是 20 世纪中格外受到关注的现象一样，无论设计创造什么样的空间，“美丽让人印象深刻”。格洛伊斯认为现当代设计让事物洗尽铅华，如果缺乏设计能够显现事物本身的主张，便无法理解设计师、艺术家、评论家与传媒人对整个 20 世纪设计艺术观创造生活的理解。为了培育并提升一介策展人的素质与自信，格洛伊斯还特别强调了策展人的地位，“策展人以公众为名，以一种公众代表的身份来管理展览空间，这个角色要捍卫它的公共特性……从源头上看，艺术品似乎是脆弱的，观众是被带到它们面前的，如同医护人员将探访者带到病人面前，策展人（curator）在词源上源自治愈（cure）并非偶然，策展就是治愈（To curate is to cure）……策展就是要治愈那些无能为力的‘图像’，因为艺术品自身无力展示自己”。从 19 世纪末到 20 世纪，艺术设计与展览历史便建立了千丝万缕的联系，瑞士艺术策展大家奥布里斯特在《策展简史》中说，20 世纪以来，策展人地位的职业化趋势变得显而易见，如很多现代艺术博物馆的创始馆长都是策展界的先驱，从 1929 年纽约现代艺术博物馆（MoMA）首任馆长阿尔弗雷德·巴尔，到 1962 年维也纳 20 世纪艺术馆（Museum de 20.Jahrhunderts/

MUMOK）的创始人维尔纳·霍夫曼都是如此。他尤其认为作为策展人要努力发掘新观念："艺术定义艺术！一切策划都要以此为基础。"

笔者及团队的建筑展览之路有十多年历程，也在理论上作过研讨——《建筑会展：为建筑师赢得话语权》（《中国建设报》（2007年12月4日）），它隶属于我主持的中国建筑文化传播项目研究之五。在我的新著《建筑传播论——我的学思片段》（天津大学出版社，2017年5月，第一版）第二章策划传播论中专论了"会展与建筑传播"问题。作为时任《建筑创作》主编，不成熟地举办过三次北京建筑设计研究院"院庆"展；2008年至2010年为创办的"中国建筑图书奖"做主题展览；2008年5月24日，与中央美术学院合作，共同为汶川"5·12"巨灾举办"痕迹艺术展"等。

近期最有价值的展览，当属2017年9月15日—17日在山东威海第九届国际人居节上举办的"致敬中国建筑经典—— 中国20世纪建筑遗产的事件·作品·人物·思想展览"。它得益于2016年9月29日，中国文物学会、中国建筑学会联合公布的首批中国20世纪建筑遗产项目。如果说历史因记录而历久弥新，建筑师的作品更因被书写而永驻人心，那么中国20世纪建筑遗产呈现的现代经典建筑及其建筑师的集体与个人史，就堪称一个国度的建筑历史"图像志"。展览经过海内外专家共同策划，围绕"致敬中国建筑经典"这一主题，在20世纪时代主线下，将事件、作品、建筑师、设计思想等要素融为一体，通过尽可能创新（实为传统模式），将中国20世纪建筑遗产项目的理念与人，亦艺术、亦文化、亦历史地展现出来，探索了用文献方法去记录中国建筑在20世纪全程流变更迭的过程，该展深得建筑业界内外的好评。

用展览讲好中国建筑文化的"故事"是个大课题，其国际视野要求我们要审视并向诸如"威尼斯双年展"学习。事实上早在1986年中国艺术家就参与了亚太博物馆策划的"开门之后：中国当代艺术展"（美国·洛杉矶），如今威尼斯双年展的影响力带动中国文化走出来。自1993年以来，由中国的艺术家、建筑师和批评家组成的"班子"不断通过威尼斯双年展向世界发声，从展览的经济投资、操作主体、操作程序等方面看，这无疑是改革发展带来的艺术展览方式的变化。展览不该是冰冷的，它一定要蕴含今天的问题，唯有精心设计与创作，唯有用文化的想象力才可召回城市的记忆。文化自信并非口号，有感于太多的"书展"是城市的文化呼吸之共鸣，让优秀展览变成有艺术技巧的可读"图书"，是中国文化走出去、吸引海外人士的标志。所以从展览思想、元素体现到互动体验，用图书展示不仅可能，而且必要。

愿在第二批“中国 20 世纪建筑遗产项目展”推出之际出版的“文化传播 · 创意设计 · 展览系列”丛书，是一个设计文化资源出新的产物，愿它带给中国建筑与文博跨界“大文化”希望。

金磊
中国文物学会 20 世纪建筑遗产委员会副会长
中国建筑学会建筑评论学术委员会副理事长
2018 年 3 月 29 日

致敬中国建筑经典
Tribute to the Chinese
Architecture Classics

致敬中国建筑经典
Tribute to the Chinese
Architecture Classics

目录

文化遗产是有生命的，这个生命充满了故事

20 世纪是人类文明进程中变化最快的时代，对中国来说，20 世纪具有特殊的意义，在 20 世纪的一百年间，我国完成了从传统农业文明到现代工业文明的历史性跨越。没有哪个历史时期能够像 20 世纪这样，慷慨地为人类提供如此丰富、生动的文化遗产。

以 20 世纪所提供的观察世界的全新视角，反思和记录 20 世纪社会发展进步的文明轨迹，发掘和确定中华民族百年艰辛探索的历史坐标，对于今天和未来都具有十分重要的意义。

首批中国 20 世纪建筑遗产项目评选认定活动，全面展示了中国 20 世纪建筑遗产项目的风采，不仅为更多的业界人士及公众领略 20 世纪建筑遗产的魅力与价值提供了重要渠道，更向世界昭示了中国 20 世纪不仅有丰富的建筑作品，也有对世界建筑界有启迪意义的建筑师及设计思想。

认真研究优秀的 20 世纪建筑遗产，思考它们与当时社会、经济、文化乃至工程技术之间的互动关系，从中吸取丰富的营养，成为当代和未来理性思考的智慧源泉。文化遗产是有生命的，这个生命充满了故事。20 世纪遗产更是承载着鲜活的故事，随着时间的流逝，故事成为历史，历史变为文化，长久地留存在人们的心中。如果说改变与创新需要智慧，那我更认为对中国 20 世纪建筑遗产保护事业要有敬畏之心，要有跨界思维，要有文化遗产服务当代社会的新策略。中国文物学会将一直支持为 20 世纪中国建筑设计思想“留痕”的工作，将与中国建筑学会合作，热情期待这项旨在保护中国城市文脉、建设人文城市之举持续开展下去。

预祝“致敬中国建筑经典——中国 20 世纪建筑遗产的事件 · 作品 · 人物 · 思想展览”成功举行！

单霁翔

中国文物学会会长
故宫博物院院长
中国文物学会 20 世纪建筑遗产委员会会长

人们所创造的建筑遗产准确地反映了那个时代

20 世纪是一个充满了重大变革、跌宕起伏的时代，在这百年间，中国的各个领域都发生了巨大的变化。无论是清末，还是国民政府时期以及中华人民共和国成立以后，人们所创造的建筑遗产都准确地反映了那个时代，反映了国家和民族的状态。随着时间的不断流逝，我们身边的建筑物成为时代的重要历史见证，其历史价值和文化价值逐渐为人们所认识。尤其是随着城市化建设的热潮，城市范围不断扩大，人口不断增长，城市发展和保护的矛盾也日渐突出。因此 20 世纪建筑遗产的认定和保护已成为一个极为紧迫的严肃课题。

建筑遗产，包括 20 世纪建筑遗产是我国文化遗产的重要组成部分，也是人类文明的重要组成部分。对建筑遗产的历史价值、文化价值、社会价值的认同是中华民族强大凝聚力的具体体现，也是自觉、自信的体现。

建筑遗产作为不可移动的物质遗产，有着不同的使用寿命、结构寿命、商业寿命、遗址历史文化寿命。在迅猛的城市化大潮中，我们往往急功近利、目光短浅，造成了不可挽回的损失和遗憾。在中国文物学会和中国建筑学会的指导下，20 世纪建筑遗产委员会将以本次遗产名录的认定为起点，和全国的有志之士一起，把关系我们的城市、城市中的人物和事件的历史记忆更好地传承下去，让这些丰富的人文内涵充实我们的民族记忆，继承伟大的民族精神。

预祝此次展览举办成功！

马国馨

中国工程院院士
全国工程勘察设计大师
中国文物学会 20 世纪建筑遗产委员会会长
北京市建筑设计研究院有限公司总建筑师

20 世纪建筑遗产是当今中国建筑师应学习的经典项目

2015 年 12 月，在中国建筑学会第十二届五次理事会暨九次常务理事会上，我在发言中提出“大建筑观”的构想，引发了与会专家的共鸣。我认为，“大建筑观”的树立，必须建立在梳理著名建筑学家的学术思想及贡献基础上。在思考中国建筑文化问题时，建筑先贤们往往从建筑全局出发，从建筑的过去、现在及未来整体走向出发，不会去人为地分割现代建筑与传统建筑。因此，他们留下的作品无论是传统的还是现代的，都堪称 20 世纪建筑遗产，都是当今中国建筑师应学习的经典项目。我们要号召广大建筑师学习传统匠人对自然的依赖与敬畏。同时，中国传统建筑文化要有现代性，要研究并分析传统建筑的价值应用，要发现何处是传统建筑的继承点，要辨识传统与现代之间的冲突点，要找到理论与实践的困惑点。中国建筑师希望创造自己的现代建筑，希望在保持中国建筑文化并吸纳西方成功建筑文化的基础上，实现中国现当代建筑的理想。

“建筑遗产”本身蕴含着文物保护与建筑设计的双重含义。中国建筑学会关注建筑遗产保护，尤其是对 20 世纪建筑遗产的保护与利用事业责无旁贷。无论是中国文物学会与中国建筑学会联合向社会公布“首届中国 20 世纪建筑遗产名录”，还是联名倡言《中国 20 世纪建筑遗产保护与利用建议书》，这些工作仅仅是双方学会合作的开端。未来，中国建筑学会将借助其深厚的专家资源与专业平台，与中国文物学会携手共同投入中国 20 世纪建筑遗产保护与利用事业，让珍贵的中国 20 世纪建筑遗产既融入人们的生活中，也为繁荣当代中国建筑创作找到可遵循的文化之根。

预祝“致敬中国建筑经典——中国 20 世纪建筑遗产的事件 · 作品 · 人物 · 思想展览”举办成功！

修龙

中国建筑学会理事长

中国建筑科技集团股份有限公司董事长

业界内外竖起 20 世纪建筑遗产的丰碑

自 2016 年 9 月 29 日，在 1914 年所建故宫宝蕴楼前广场上，由中国文物学会、中国建筑学会联合公布首批共 98 项“中国 20 世纪建筑遗产项目”以来，业界内外反响热烈，中国现代经典建筑标志的呈现不仅记录着 20 世纪的建筑史，也丰富着 20 世纪的事件史。历史因记录而历久弥新，建筑师的作品更因被书写而永驻人心，这是建筑的国家历史“图像志”。

20 世纪是中西方文化激烈碰撞的时代。城市建筑、文博史论界专家在感悟这百年经典时认同：它是中华人民共和国历史上第一次由中国文物学会、中国建筑学会联手，在创造卓越的“国家记忆”作品的推介上迈出的一步；《中国 20 世纪建筑遗产名录（第一卷）》在全面梳理了 98 项成果脉络、展示中国建筑经典整体风貌的同时，还写就了 20 世纪中国建筑师集体与个人的建筑史；入选作品体现了时代背景下中国建筑学家坚守信念与创新的使命感，在业界内外竖起 20 世纪建筑遗产的丰碑，有理由说，中国建筑已成为 20 世纪世界建筑家族中的一种风格与流派，向世界打开以建筑名义看中国的“窗口”；20 世纪中国建筑与国家叙事有天然的联系，如果人类的文明需要新故事支撑，相信大家会从中读懂敬畏中国建筑经典的文化自尊与自信。

本展览围绕“致敬中国建筑经典”的主题，在贯穿 20 世纪时代主线的背景下，将事件、作品、建筑师、设计思想等要素融为一体，让观众通过这个亦建筑、亦艺术、亦文化的展览，理解经典建筑的“出炉”经历了哪些过程。本展览是一个用文献方法记录中国建筑在 20 世纪流变更迭的“工具”，它是照亮城市文化的一盏盏灯，更是国家的文明使者与当代标志。愿在第二批“中国 20 世纪建筑遗产名录”揭幕前夕举办的本展览，是跨文化设计资源的探索，是建筑创作的传统出新，是体现中国建筑意境与当代视野的“文化工程”，也希望它在传播建筑界、文博界理念的同时，为“重读”中国现当代建筑设计史带来新的谱系。

金磊

中国文物学会 20 世纪建筑遗产委员会副会长、秘书长

《中国建筑文化遗产》《建筑评论》主编

西风渐进

Gradual Spread of Western Influences

1900–1928

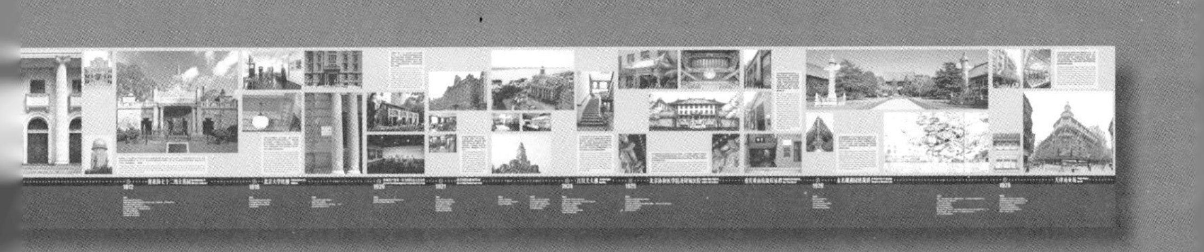

20世纪初叶是中国现代建筑崛起的年代，这恰好也是西方现代建筑步入成熟的时期。在这一时期的许多中国城市中，大型建筑设计项目主要由外国人把持——建筑变革随西方的思想、文化进入中国，外国建筑师将本国的建筑设计思想和方法带到中国，对中国现代建筑的发展起到了巨大的影响。邬达克、墨菲、司马等外国建筑师的到来， 马可·波罗广场、徐家汇教堂、上海外滩、北京协和医科大学及医院等建筑的出现，为当时的中国带来了从未有过的设计变化。中国建筑处于被动的输入和主动的发展状态。

The early decades of the 20th century saw the rise of modern Chinese architecture. It was exactly in the same period that western modern architecture became mature. In a number of Chinese cities in this period, large architectural design projects were mainly conducted by foreigners. With the introduction of western ideas and culture, architectural revolution took place in China. Foreign architects brought their architectural design ideas and methods into China, which imposed huge influences on the development of modern Chinese architecture. The arrival of such foreign architects as László Hudec, Henry Murphy and Small and emergence of such buildings as Marco Polo Square, Xujiahui Cathedral, the Bund of Shanghai, Peking Union Medical College and Hospital had brought changes in the architectural design that had never been known before in China then. Chinese architecture was in a state of passive input and active development.

佘山天文台
Sheshan Observatory

1900

北京、保定等多处古城及古建筑群，北京西郊清漪园等行宫遭到八国联军破坏和劫掠。
旅顺火车站建成。
上海泰兴洋行大班住宅建成。
由俄国皇太子倡议的汉口东正教堂建成。
德国建筑师制定青岛城市规划。
建筑学家童寯（1900—1983）出生。

佘山天文台建于清光绪二十六年（1900 年），为法国传教士所建造的具有欧洲建筑风格的天文台，占地 8000 余平方米，是我国最早的现代意义上的天文台，也是我国的天文研究中心之一。它拥有当年“远东第一”的 40 厘米双筒折射望远镜，百年来拍下了 7000 多张珍贵的天文照片。其 2013 年被列入第七批全国重点文物保护单位。

The Sheshan Observatory in European style was built by a French missionary in 1900, the 26th year of the Reign of Emperor Guangxu in Qing Dynasty. Covering an area of 8,000 square meters, it is the earliest modern observatory and one of the astronomical research center in China. It was then equipped with the "Far East Best" 40 centimeters binocular refractor that took over 7,000 valuable astronomy pictures of the past century. In 2013, the observatory was listed among the 7th Batch of Important Historical Sites under State Protection.

主入口 / 室内展室旧影

1901

法国修建的越南河内至中国广西龙州铁路竣工。
上海第一座东正教堂在上海闸北兴建。
建筑学家梁思成（1901—1972）、杨廷宝（1901—1982）出生。
北京瑞福祥建成。

1902

英国教会设新书院于天津。
汉口俄国领事馆建成。
保定陆军军官学校建成。
哈尔滨华俄道胜银行建成。

1903

清廷再次重修颐和园。
北京美国公使馆办公楼建成。
庐山巴雷夫人别墅建成。
京汉铁路南端终点建筑大智门火车站建成。
旅顺大狱建成。

北洋大学堂旧址

Site of Peiyang University

北洋大学是中国近代史上的第一所大学，成立于 1985 年。1951 年 9 月与河北工学院合并，定名为天津大学，校园旧址在今河北工业大学校园内。北洋大学旧址现存有南大楼、北大楼、团城三座建筑，南大楼、北大楼均为三层砖混结构的楼房，建筑布局对称，体形简洁大方，外立面为红砖墙面。

Peiyang University, which established in 1895, is the first university in Chinese modern history. It merged with Heibei Institute of Technology and named Tianjin University in 1951. The site of Campus is in today's Hebei University of Technology. The site of Peiyang University now consists of South Building, North Building, and Tuancheng. The South Building and North Building have the same simple symmetrical three-storied structure of brick and concrete. Each building has its front facade in red bricks.

西泠印社

Xiling Seal Art Society

西泠印社位于杭州西湖边，是中国研究金石篆刻的一个百年学术团体，有“天下第一名社”之称。西泠印社坐落于杭州市西湖景区孤山南麓，占地面积 7088 平方米。西泠印社居山而建，由上、中、下三部分组成，各具特色，建筑与周围环境融为一体。

Lying by the West Lake in Hangzhou, Xiling Seal Art Society is a century-old academic society studying inscription and seal cutting in China, which has been known as the Most Famous Society in China. Covering an area of 7,088 square meters, Xiling Seal Art Society sits on the south of Gushan Hill of West Lake spot in Hangzhou. Built on the hill slope, the complex consists of three sections, i.e. the upper, middle and lower sections, each with distinctive features but all in harmony with the surroundings.

华严经塔 / 西泠印社旧影 / 1988 年绘《西泠印社图》

1904

日本在长沙小西门外修建湖南轮船公司码头。

青岛火车站建成。

北京宣武门天主堂第六次重建。

汉口美国领事馆建成。

京汉铁路竣工。

清廷批准并颁布《奏定学堂章程》，为中国教育史上第一个正式颁布并在全国普遍实行的学制；此章程规定大学堂分为八科，在工科下设置建筑工学和土木工学。

哈尔滨火车站建成。

北京王府井大主教堂重建。

1905

詹天佑主持筹建京张铁路。

上海建法租界公董局大楼。

胶济铁路及大港码头建成。

宁夏董府建成。

成都平安桥天主教堂建成。

天津英国俱乐部建成。

厂区鸟瞰 / 厂房内景

1906

哈尔滨中东铁路管理局大楼建成。
天津劝业会场落成。
青岛提督府建成。
京奉铁路正阳门东车站开始建设。
哈尔滨铁路技术学校建成。
哈尔滨莫斯科商场建成。

1907

陕西筹建延长石油厂。
北京中华圣公会教堂建成。
北京陆军部南楼建成。
天津广东会馆建成。
保定育德中学建成。

华新水泥厂旧址
Site of Huaxin Cement Factory

华新水泥厂旧址位于湖北省黄石市，是中国现存生产时间最长、保存最完整的水泥工业遗存。华新水泥厂原名大冶湖北水泥厂，是中国近代较早开办的水泥厂之一，创建于清光绪三十三年（1907 年）。华新水泥厂旧址整体保存完整，见证了中国民族工业从萌芽、发展到走向现代的历史进程。

The site of Huaxin Cement Factory in Huangshi City of Hubei is the most well-preserved cement industry relic existing with the longest history. Built in 1907 or the 33rd year of the Reign of Emperor Guangxu in Qing Dynasty, it is one of the earliest cement factories of China. Huaxin Cement Factory witnessed the history of Chinese national industry from sprouting, growing and leaping into modernity.

建筑立面局部 / 礼拜堂内饰

1908

津浦铁路济南火车站建成。此建筑后于1992年被拆除。
北京农事试验场附设公园开放。
上海大北电报公司大楼建成。
德国驻济南领事馆建成。

1909

北京资政院大楼落成。
昆明云南陆军讲武学堂建成。
南通市着手城市规划。
中国近代第一座公共图书馆——武昌文华大学图书馆落成。
《大连市区规划》编制完成。
北京女子师范学堂建成。

徐家汇天主堂
Xujiahui Cathedral

徐家汇天主堂是中国著名的天主教堂，为天主教上海教区主教座堂。建筑风格为中世纪哥特式。这座主教座堂于清光绪三十年（1904 年）动土兴建，清宣统二年（1910 年）10 月落成。建筑平面布局呈“T”形，大门朝东，为砖木结构，采用法国中世纪样式。

Xujiahui Cathedral is a famous cathedral in China. The buildings adopt a medieval Gothic architectural style. The cathedral started to be built in 1904 (the 30th year of the Reign of Emperor Guangxu in the Qing Dynasty) and was completed in 1910. With a T-shaped plan, the entire building is a French medieval-style structure of brick and timber whose main gate faces east.

1910

在上海开业的外国建筑师或合伙事务所已达14家。此后，英国建筑师威尔逊于1912年在上海建立公和洋行，美国建筑师哈沙德经管哈沙德洋行。后，南京国民政府聘请美国建筑师墨菲和匈牙利建筑师邬达克为城市建设顾问专家。

滇越铁路建成。

湖北省咨议局大楼、江苏省咨议局大楼、广东省咨议局大楼建成。

北京外务部迎宾馆建成。

津浦铁路天津西站建成。

上海徐家汇天主堂建成。

北京亚斯立堂建成。

汉口最早的多层公寓巴公房子建成。

武昌起义军政府旧址

Site of the Government of Wuchang Uprising

武昌起义军政府旧址（湖北军政府旧址、鄂军都督府旧址）位于武汉长江大桥武昌引桥东端，紧靠蛇山南麓，其前身为清末“湖北省咨议局”机关大楼，始建于清光绪三十四年(1908年)，清宣统二年（1910年）建成，由武昌起义军政府旧址主楼（咨议局办公大楼）、东西配房、议员公所、前后花园、大门、门房及围墙等建筑组成。旧址主楼为西式日本议院建筑风格的砖木结构的二层楼房。主楼平面呈“山”字形，为砖混结构，墙体用特制的优质红砖砌成清水墙，故该建筑也有“红楼”之称。

The Site of the Government of Wuchang Uprising formerly is the building of the “Hubei Provincial Assembly” in late Qing Dynasty. The construction was finished in the 2nd year of the Reign of Emperor Xuantong in the Qing Dynasty (the year 1910). The site consists of main building of the Government of Wuchang Uprising (office building of the Provincial Assembly), east and west wings, senator’s office, gardens in the forecourt and backyard, main gate, gatehouse and enclosing walls. The wall is made into dry wall with special qualified bricks, thus, it’s also called “red building”.

建筑主入口

清华大礼堂立面 / 清华大礼堂柱础 / 清华大礼堂柱头

清华大学早期建筑

Early Architecture Of Tsinghua University

1911

美国教会在杭州筹建之江大学。
北京开始营建清华学堂清华园建筑群。

目前所存清华大学早期建筑主要有三部分：1911—1912 年建造的清华学堂、同方等；1919—1925 年建造的大礼堂、图书馆（局部）等；1931—1933 年建造的生物馆、气象台、校门等。清华大学早期建筑整体保存较好，使校园保留了近代校园的典雅风格，至今仍在为教学和科研服务。

The existing Tsinghua University's early architecture consists of three parts: the Tsinghua College and Tsinghua Tongfang built during 1911–1912; the auditorium and library (partial) built during 1919 1925; and, Hall of Biology, meteorological station, main gate, etc. These buildings constructed during Tsinghua University's early years are well preserved and blended with the modern buildings on the campus, emanating a stylish sensation to serve the teaching and research activities continuously.

黄花岗七十二烈士墓园

Huanghuagang 72 Martyrs' Cemetery

黄花岗七十二烈士墓园位于广州市北面的白云山南麓先烈中路，是为纪念 1911 年 4 月 27 日（农历辛亥年三月二十九日）孙中山先生领导的同盟会在广州“三二九”起义战役中牺牲的烈士而建的。1921 年，烈士墓和纪功坊先后落成。墓园占地 16 万平方米，建筑规模宏大，气魄雄伟。

Huanghuagang 72 Martyrs' Cemetery, located on Xianlie Middle Road, south of Baiyun Mountain in the north of Guangzhou City, was built to commemorate the martyrs of Tung Meng Hui (Chinese revolutionary league) led by Mr. Sun Yat-sen in the March 29th Uprising in Guangzhou on April 27, 1911. The Mausoleum of the Martyrs and merit archway were completed in succession in 1921. Covering an area of 160,000 square meters, the cemetery is grand in scale and magnificent.

纪念碑正立面及环境 / 墓道旧影

1912

津浦铁路洛口黄河大桥竣工。
南京原两江总督署改建为临时大总统办公地（后由姚彬、虞炳烈等设计，扩建为南京国民政府总统府）。
南京为杨卓林、郑先声等烈士建专祠。
杭州将清万寿宫改为“南京阵亡将士祠”。
西安易俗社剧场建成。
昆明石龙坝水电站建成。
北京国会旧址开始建设。

1913

哈尔滨马迭尔饭店落成。
成都辛亥秋保路死事纪念碑建成。
上海建福新面粉厂。
陈嘉庚创办厦门集美学村。
上海杨树浦发电厂建成。

北京大学图书馆
Peking University Library

北京大学图书馆前身为建立于1898年的京师大学堂藏书楼，辛亥革命后改名为北京大学图书馆。1952年，原燕京大学图书馆与北京大学图书馆合并。北京大学图书馆老馆是美国建筑师墨菲于1920年规划设计的，是当时燕京大学建筑群中的主要建筑之一。20世纪70年代和90年代，对其先后进行了扩建。

Renamed after the Xinhai Revolution, Peking University Library is once the library of the former Imperial University of Peking, which was established in 1898. In 1952, the library merged with the former Yenching University Library. The old library, designed by American architect Murphy in 1920, was one of the buildings of the former Yenching University. In the 1970s and 1990s, the library went through two major extensions.

图书馆旧馆一 / 图书馆旧馆二 / 全景鸟瞰

马可·波罗广场全景／民族路43号／自由道38号

1914

天津开始建造“天津渤化永利碱业有限公司”。
上海建成兆丰公园（1925年后命名为“中山公园”）。
北京改建社稷坛为中央公园（后命名为“中山公园”）。
南通濠南别业落成。
大连大和饭店建成。
天津梁启超住宅建成。

1915

江西南浔铁路竣工。
朱启钤（时任北洋政府内务总长）主持北京旧城改建工程。
胶济铁路济南火车站建成。
台湾总督府建成。

1916

天津西开教堂建成。
广州西堤邮电局建成。
长沙湘雅医学院建成，规划设计者为美国建筑师墨菲。

马可·波罗广场建筑群

Architectural Complex around Marco Polo Square

马可·波罗广场是随着20世纪初意租界规划开辟而形成的，占地2200平方米。广场周边从1908年至1916年建成了意式花园别墅住宅建筑群。建筑群由和平女神雕塑和6座典型的意大利南部地中海风格别墅组成。

Marco Polo Square was built with the development of the Italian Concession in the early 20th century. Covering an area of 2,200 square meters, the Italian-style garden villa compound around the square was built between 1908 and 1916. The architectural complex on the square consists of a sculpture set and six villas in Mediterranean style of southern Italy.

汉口近代建筑群

Modern Architectural Complex of Hankou

汉口近代建筑群位于武汉市汉口旧大智门火车站以南至江汉路，京汉大道以东至长江边，包括汉口的英国、法国、德国、俄国、美国领事馆旧址，汉口日清洋行，汉口横滨正金银行，汉口花旗银行，武汉港，江汉关监督公署等。汉口近代建筑群是西式风格建筑或中西结合形成的新式建筑，有着鲜明的时代特点。

The modern architectural complex of Hankou in Wuhan is located between the south of the old Dazhimen Railway Station and Jianghan Road and the east of Jinghan Avenue to the bank of the Yangtze River, consisting of the site of the British, French, German, Russian, and American consulates in Hankou, as well as Nissin Foreign Firm in Hankou, Yokohama Specie Bank in Hankou, Citibank in Hankou, Wuhan Port and Jianghan Customs. The buildings are either Western-style or an integration of Chinese-Western architectural style, reflecting vivid characteristics of the times.

长沙岳麓山黄兴、蔡锷墓建成并举行国葬。
上海先施公司建成。
芜湖中国银行建成。
鼓浪屿天主堂建成。

汉口东正教堂 / 汉口俄国领事馆 / 八七会议旧址

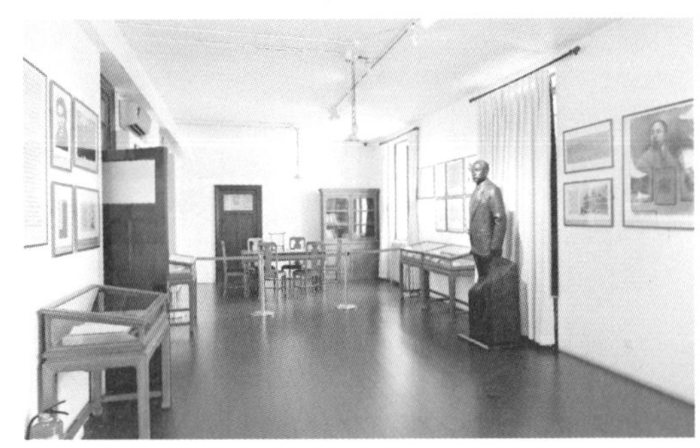

正立面主入口 / 室内展厅

北京大学红楼
Peking University Honglou

北京大学红楼简称“北大红楼”，是 1916 年至 1952 年期间北京大学的主要校舍之一。建筑通体用红砖砌筑，红瓦铺顶，故名“红楼”。北大红楼于 1918 年落成，为砖木结构建筑，东西宽 100 米，正楼南北进深 14 米，总面积约 1 万平方米，现为北京新文化运动纪念馆。

Peking University Honglou, or Beida Honglou for short, was one of the major buildings of Peking University during 1916–1952. The entire building is constructed all over with red bricks and roofed with red tiles, hence it's named Honglou.Peking University Honglou, completed in 1918. Honglou, a four-storied building of brick and timber, is 100 meters wide from east to west; the main structure is 14 meters deep from north to south. The entire floor area is approximately 10,000 square meters. Honglou is Beijing New Culture Movement Memorial Hall now.

1918

由蔡元培倡导创立北京美术学校。
广州营建黄花岗七十二烈士墓园。
旅顺俄国陆军将校俱乐部建成。
天津张园建成。

1919

上海公共租界工部局大楼建成。
金陵大学北大楼建成。
朱启钤主持重新刊行宋《营造法式》（丁本）。
武汉德明饭店建成并开业。

入口及外景／室内展陈／立面细部

中国共产党第一次全国代表大会会址

Site of the First National Congress of the Communist Party of China

中国共产党第一次全国代表大会（简称“一大”）会址位于上海市兴业路76号。“一大”原址纪念馆是两栋砖木结构的二层石库门楼房，坐南朝北，属典型的上海石库门式样建筑。其外墙青红砖交错，镶嵌着白色粉线，门楣有红色雕花，黑漆大门上配铜环，门框围以米色石条。1949年后对原址进行了修复。

The site of the First National Congress of the Communist Party of China (CPC) was at No. 76, Xingye Road, Shanghai. The two two-storied north-facing brick-wood shikumen-style buildings constitute today's memorial hall. The external walls are of interlaced red and blue bricks inlaid with white-plaster lines. The architrave is of red-color carved flowers. On the black-lacquered gate, there are two bronze ring knockers and a cream-colored stone frame. In 1949, a renovation was made to restore the meeting site.

1920

美国建筑师墨菲开始规划设计北京燕京大学。
关颂声在天津创办基泰工程司。

建筑沿街立面 / 室内展室

武汉国民政府旧址

Site of National Government Office

武汉国民政府旧址位于武汉市汉口中山大道，该楼于1917年兴建，为钢筋水泥结构，坚固宏伟，富丽典雅。其中部6层，两侧为5层，对称布局，气势雄伟。建筑占地885平方米，建筑面积为6740平方米。

The site of National Government Office in Hankou's Zhongshan Avenue of Wuhan was built in 1917. The architecture is of reinforced concrete, hard, masculine, splendid, and elegant. The central section is six-storied and the two symmetrical wings are five-storied. The entire building is symmetrical and grand. The site has an area of 885 square meters, and buildings with an area of 6,740 square meters.

1921

天津开滦矿务局大楼落成。
北京协和医学院建筑群落成。
汉口景明洋行大楼建成。
陈嘉庚创办厦门大学。
北京真光电影剧场建成。
云南德钦茨中教堂落成。

江汉关大楼

Jianghanguan Building

江汉关大楼位于武汉市沿江大道与江汉路交会处，占地 1499 平方米，主楼四层，底层为半地下室，钟楼四层，总高度 46.3 米，为武汉当时最高的建筑，于 1924 年落成。建筑外墙用花岗石砌成，东、西、北三面墙均带花岗石廊柱，饰以变形的科林斯柱头式样。立面采用三段构图的表现手法，处理手法体现出文艺复兴时期的建筑风格。

Situated at the intersection of Yanjiang Avenue and Jianghan Road in Wuhan, the 46.3-meter high Jianghanguan Building was built in 1924, which was the highest building at that time in Wuhan.The exterior walls are of granite and the granite columns are ornamented with Corinthian variant chapiters on the east, west and north. The front facade adopts a three-section composition. The way of handling reflects the Renaissance architectural style.

1922

金陵女子学院建成。
救世军中央堂旧址开始建设。
天津平安影院建成。

1923

汇丰银行上海分行大楼建成。
哈尔滨土耳其清真寺建成。
近代中国第一所建筑院校——苏州工业专门学校成立。
天津国民饭店建成。

1924

上海建造霞飞坊等新式里弄住宅。
上海宋公园（今闸北公园）树立宋教仁石雕坐像。
武昌首义纪念公园建成开放。
汉口江汉关大楼建成。
上海大世界游乐场建成。
云南弥勒烂泥箐天主教堂建成。
云南大学会泽楼建成。

江汉关旧影

北京协和医学院及附属医院

Peking Union Medical College and Hospital

北京协和医院位于北京城市中心的王府井地区，1921年落成。建筑群以门诊楼为中心，沿两条相互垂直的轴线对称布置，西门入口处为歇山顶抱厦门廊，大红柱身，彩绘梁枋。虽然中国传统建筑的要素在建筑中到处有所体现，但为满足使用要求而进行的调整也有许多。

Peking Union Medical College and Hospital is located in the city center of Beijing Wangfujing area, completed in 1921. Buildings to the outpatient building as the center, along the two vertical axis symmetrical layout, west entrance for the Gablet roof Gallery, the red column body, painted Liang Fang. Although the elements of traditional Chinese architecture are reflected everywhere in the building, there are many adjustments to meet the requirements of the use.

1925

刘既漂设计巴黎万国博览会中国馆。
上海佘山大教堂奠基。
孙中山先生葬事筹备处征集陵墓建筑图案，吕彦直设计方案获头奖。
上海新新公司大楼建成。
云南大理天主堂建成。
故宫博物院成立。

斗拱局部／入口装饰／立面手绘图

重庆黄山抗战旧址群

Chongqing Huangshan War Site Cluster

抗战时期国民政府迁至重庆后，以重金买下该宅邸，供蒋介石及国民政府要员和宋氏家族、孔氏家族等人居住。旧址有云岫楼、松厅、孔园、云峰楼、松籁阁及美军顾问团居住地莲青楼、马歇尔住地草亭和抗战遗孤学校黄山小学等 13 处房屋，是国内抗战遗址中最大、保存最完整的遗址群。

After moving to Chongqing during the War of Resistance against Japanese Aggression, the National Government bought the residence offered as living quarters for Chiang Kai-shek, senior officals of the National Government, and family members of the Songs and Kongs. The residence consists of 13 living buildings, including Yunxiu Building, Songting Hall, Kongyuan Garden, Yunfeng Building, and Songlai Pavilion, and Lianqing Building for the US Military Assistance Advisory Group, Caoting Hall for General Marshall and Huangshan Elementary School for orphans of victims in the Resistance War. It is the largest and best preserved historical sites.

云岫楼侧立面 / 云岫楼入口

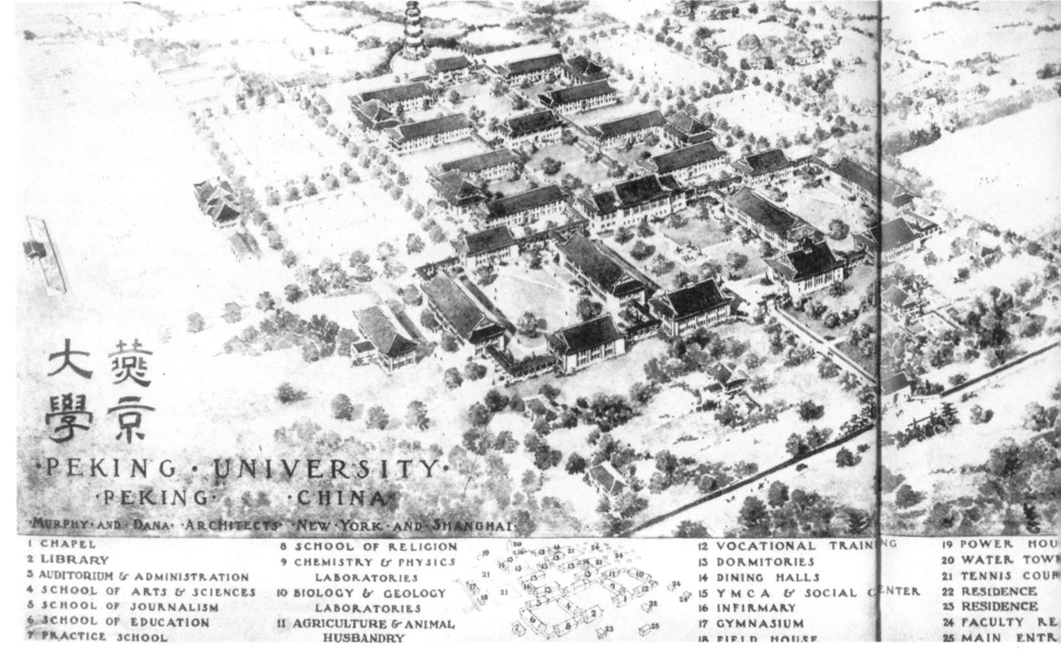

燕京大学早期规划图

未名湖燕园建筑群

Architectural Complex of Yanyuan Campus by Weiming Lake

未名湖燕园建筑群位于北京市海淀区北京大学校园内，是近代著名学府燕京大学原址。燕京大学以明代名园“勺园”故址为中心兴建校舍，1926 年建成。该建筑群以未名湖为中心，分布于四周，全部为仿古建筑。各群组建筑都为三合院式，总体布局合理，局部尺度适宜，与自然地形地貌结合紧密。

Yanyuan Campus by Weiming Lake of Peking University is in Haidian District of Beijing. It is the site of Yenching University. Yenching University was built around the site of Shaoyuan Garden, a known garden of the Ming Dynasty. Its construction completed in 1926. The architectural complex is centered on Weiming Lake. All buildings are in imitation of ancient architectural style. All architectural groups are in sanheyuan formation (a courtyard enclosed on three sides). The overall layout is reasonable, well proportioned, and blended with the surrounding natural environment.

1926

梧州中山纪念堂奠基。
天津万国桥建成。
盐业银行天津分行大楼建成。
汉口万国跑马场建成。
中华圣经会旧址开始建设。

1927

庄俊、范文照倡议成立上海建筑师学会，次年更名为中国建筑师学会。
上海市成立都市计划委员会。
长沙中山纪念堂建成。
上海江海关大楼建成。
厦门中山公园建成开放。
国立第四中山大学成立，其中建筑科由苏州工业专门学校建筑科并入，并经过重新组建。至此，中国第一个大学建筑系正式成立。
东北大戏院在沈阳建成。

檐口及窗框装饰 / 东北立面

1928

上海英商正广和汽水厂大班住宅建成。
美国建筑师开尔思设计规划国立武汉大学。
东北大学设建筑系，梁思成任系主任。
汉口中山公园建成开放。
广东汕头中山公园建成开放。
中央古物管理委员会在南京成立。
沈阳“大帅府”建成。

天津劝业场

Tianjin Quanye Bazaar

天津劝业场位于天津市和平区滨江道的商业中心区。天津劝业场的建筑风格明显受折中主义建筑形式的影响。其主体五层，转角局部七层，为钢筋混凝土框架结构。七层之上建有高耸的塔楼，由两层六角形的塔座、两层圆形的塔身和穹隆式的塔顶组成，上面装有旗杆、避雷针兼作装饰物。整栋建筑显得壮丽挺拔。

Tianjin Quanye Bazaar, located in the central business area by Binjiang Road, Heping District, Tianjin City.In terms of architectural style, Tianjin Quanye Bazaar has obviously been influenced by eclectic architecturc. Adopting thc rcinforcod concrete frame structure, it consists of the five-storied main constructions and seven-storied constructions at the corners. On the seven-storied constructions, there are high towers, each of which is made up of a two-storied octagonal base, a two-storied round body and a dome-style roof decorated with banner poles and lightning rods. The entire building looks tall and splendid.

中国固有之形式

Inherent Forms of China

1929–1949

1929年南京国民政府的《首都计划》以及同时期的《上海市中心区域规划》都体现了中国建筑“文化本位”的思想，从而提出“中国固有之形式”。加之中国建筑师对中国建筑的探索，形成了一股与外国建筑师相抗衡的建筑设计力量，在建筑民族化、建筑与历史文化、地域文化相结合等方面做出了前无古人的突出贡献。庄俊、吕彦直、柳士英、陈植、童寯等建筑师的表现，以及中山陵、中山纪念堂、国立紫金山天文台、南岳忠烈祠、黄花岗七十二烈士墓园、南京国民政府建筑群经典项目的涌现，标志着中国建筑师的崛起和成长。他们将中国传统建筑与西方现代建筑思想相结合，其作品不仅带有中国建筑的种种印痕，还展示了超常的技巧与能力，为中国现代建筑及新民族形式的发展进行了有益的实践。

Both the *Capital Plan* in 1929 in the reign of Nanjing National Government and *Downtown Area Plan* of Shanghai during the same period reflected the "culture-based" idea of Chinese architecture, and thus the "inherent forms of China" was put forward, forming a force of architectural design which could compete against foreign architects, combined with Chinese architects' exploration into the Chinese architecture. Unprecedented outstanding contributions had been made to the combination of nationalized architecture, architecture and historical culture and regional culture. The achievements of Zhuang Jun, Lv Yanzhi, Liu Shiying, Chen Zhi, Tong Jun and other architects and the emergence of typical projects such as the Sun Yat-sen Mausoleum, Sun Yat-sen Memorial Hall, Zijinshan National Astronomical Observatory, Nanyue Martyrs' Shrine, Huanghuagang 72 Martyrs' Cemetery and Nanjing National Government Building Relics marked the rise and growth of Chinese architects. They combined traditional Chinese buildings with western modern architectural ideas, with both traces of Chinese buildings and excellent skills and capabilities, providing beneficial practices for the development of modern Chinese architecture and new national forms.

中山陵

Sun Yat-sen Mausoleum

中山陵是中国近代伟大的民主革命先驱孙中山先生的陵寝及其附属纪念建筑群，亦是“中国固有式建筑”的开山之作。整个建筑群依山势而建，由南向北沿中轴线逐渐升高。主要建筑有博爱坊、墓道、陵门、石阶、碑亭、祭堂和墓室等，排列在中轴线上，体现了中国传统建筑的风格。

Sun Yat-sen Mausoleum is the mausoleum and the appended memorial buildings of Mr. Sun Yat-sen, great forerunner of the Chinese democratic revolution in modern times. The building complex is situated on, and adapted to, the hillside. Rising gradually from south to north along the middle axis, it consists primarily of the Memorial Archway of Universal Fraternity (Bo'aifang), tomb avenue, mausoleum gate, stone steps, Marble Pavilion (Beiting), Sacrificial Hall (Jitang), and tomb chamber, which arranged along the middle axis, shows a typical style of Chinese traditional architecture.

鸟瞰图 / 祭堂及墓室

1929

朱启钤自筹资金在北平成立营造学社，自任社长（1930年更名为中国营造学社，梁思成、刘敦桢分别为法式部、文献部主任，所创办《中国营造学社汇刊》刊载调查报告等研究成果，共出版七卷23期）。

刘既漂受委托主编《旅行杂志》建筑专号，系统介绍了西方建筑概念。

南京中山陵建成并举行奉安大典。

南京公布《首都计划》。

广州海珠桥建成。

上海沙逊大厦建成。

梁思成设计吉林大学教学楼。

浙江昭忠祠改建为浙江忠烈祠。

宁波中山公园建成开放。

董大酉等编制完成《上海市中心区规划》。

入口门楼 / 原司法院及司法行政部旧影

孙中山临时大总统府及南京国民政府建筑遗存

Sun Yat-sen Provisional Presidential Palace & Nanjing National Government Building Relics

该建筑为清朝时期两江总督端方创意建造的一座坐北朝南的西式平房建筑，面阔 7 间，中间有设计精巧的亭形拱式门斗。南京市区还建有“行政院”“外交部”“交通部”等国民政府建筑，大多由中外知名建筑师（如杨廷宝、童寯、墨菲等）设计，凝聚了中国近代建筑之精华。

This building is a south-facing, western-style one-storied house built at the request of Duan Fang, governor-general of Liangjiang during the Qing Dynasty. Seven bays wide, it has a pavilion-like arched porch in the middle. The urban area of Nanjing also boasts many other buildings that used to house various departments of the National Government, such as the Executive Yuan, the Foreign Ministry, the Ministry of Transport. Designed by famous architects from home and abroad (e.g. Yang Tingbao, Tong Jun and Henry Murphy), they represent the quintessence of modern Chinese architecture.

天津五大道建筑群

Architectural Complex in Five Great Avenues of Tianjin

1930

南京铁道部、交通部等政府办公建筑相继建成。
哈尔滨交通银行建成。
云南大理天主教堂建成。
北京辅仁大学建成。
天津东莱银行建成。
京奉铁路沈阳总站建成。

五大道在天津市城区的南部，沿东西向并列着以重庆、大理、常德、睦南及马场为名的五条街道，因此被称作“五大道”。五大道地区拥有20世纪二三十年代建成的具有不同国家建筑风格的花园式房屋2000多所，建筑面积达100多万平方米，被称为万国建筑博览会。

Five Great Avenues, in the south of Tianjin, gain its name from five east-west parallel avenues, i.e. Chongqing Avenue, Dali Avenue, Changde Avenue, Munan Avenue, and Machang Avenue. In the area, there are over 2,000 garden houses of different exotic architectural styles built during the 1920s-1930s. The gross floor area is over 1,000,000 square meters, which are called International Architecture Expo Garden.

鸟瞰 / 成都道与长沙道交口

国民政府行政院旧址
Site of Executive Yuan of National Government

1930

国民政府行政院旧址坐落于南京市中山北路252号。1930年竣工后始为原国民政府铁道部办公楼，1945年原国民政府行政院迁入。原国民政府行政院旧址建筑由一组大屋顶传统宫殿式建筑组合而成。主体为三层，两侧为二层，北端为四层，另有一层地下室，总建筑面积9466平方米。

The original site of the Executive Yuan of National Government is located in No.252 Zhongshan North Road, Nanjing. After the completion in 1930, it became the former National Government Ministry of Railways office building, then in 1945 the original National Government Executive Yuan moved in. The building embodied the "Chinese inherent form" under the domination of the "Chinese standard".The main body is of the three-story, on both sides of the two-story, the northern end of four-story, and another layer of basement. The total construction area is 9,466 square meters.

建筑内景 / 外立面旧影

外立面及环境 / 建设过程 / 剖面图 / 东立面图

中山纪念堂

Sun Yat-sen Memorial Hall

1931

广州市中山纪念堂建成。
长沙中山亭（钟楼）建成。
上海国际饭店奠基（1934年建成）。
国立北平图书馆落成。
天津法租界工部局大楼建成。
广州市府合署落成。
中央银行建成。

广州市中山纪念堂是“中国固有式建筑”的代表作之一，为 1921 年孙中山先生在广州出任临时大总统时的总统府旧址。中山纪念堂是一座八角形的宫殿式建筑，正立面为七开间朱红色柱廊，庄严瑰丽。堂内有一个近似圆形的大会堂，上下两层，共有 6000 多个座位。

The Sun Yat-sen Memorial Hall is one of the representative works of "Chinese inherent architecture". It was formerly the presidential palace for Mr. Sun Yat-sen when he was Interim President in Guangzhou in 1921. Sun Yat-sen Memorial Hall is a dignified and splendid octagonal palatial architecture featuring a seven-bay facade with vermilion colonnade. Inside the hall, there is a two-storied quasi-circular conference hall with more than 6,000 seats.

图书馆三期立面 / 图书馆二期入口 / 二期外立面

清华大学图书馆
Tsinghua University Library

清华大学图书馆前身为 1911 年 4 月创办的清华学堂图书室。1919 年 3 月，现老馆东部落成，建筑面积为 2114 平方米，迁入新馆舍的同时，更名为清华学校图书馆。1930 年 3 月开工扩建馆舍（今老馆中部和西部），馆舍面积增至 7700 平方米。1991 年 9 月，由香港邵逸夫先生捐资和国家教委拨款兴建的新馆落成，被命名为“逸夫馆”。

Tsinghua University Library was initially the library of Tsinghua College. The east section of the old venue was completed in March, 1919, with a floor area of 2,114 square meters. In March, 1930, the old venue started to be expanded (the middle and west sections of the old venue), with the floor area increased to 7,700 square meters. In September, 1991, the new venue jointly funded by Sir Run Run Shaw of Hong Kong and the National Education Commission was completed. Later it was renamed Run Run Shaw Building.

建筑沿街立面 / 建筑外景

国民参政会旧址

Site of National Political Council

国民参政会是抗战时期由国共两党及其他党派、无党派人士代表组成的最高咨询机关，是一个具有广泛政治影响的议会机构。国民参政会旧址位于重庆市渝中区中华路，是一座别致的西式小楼，二楼一底，宽18.9米、进深193米、高15.3米，共有房屋21间。

National Political Council was the highest and the most influential council assembly of representatives from the Kuomintang (KMT), Communist Party of China (CPC) and other parties as well as nonparty personages. The site was on Zhonghua Road, Yuzhong District, Chongqing, in a unique three-storied Western style, 18.9 meters wide, 193 meters deep, 15.3 meters high, and partitioned into 21 rooms.

1932

哈尔滨圣·索菲亚教堂建成。

中国建筑师学会主办《中国建筑》月刊刊行。

上海建筑师协会主办《建筑月刊》杂志创刊。

赵深、陈植、童寯在上海创办华盖建筑师事务所。

日伪当局在长春实施“新京国度建设计划”。

圣·索菲亚教堂

Saint Sophia Cathedral

圣·索菲亚教堂坐落于哈尔滨市道里区，1932年11月落成。该教堂为拜占庭风格的东正教教堂。建筑高53.35米，占地721平方米。屋顶有巨大的“洋葱头穹顶”，四翼的建筑屋顶有大小不同的“帐篷顶”，错落有致。四个楼之间有楼梯相连。建筑平面呈希腊十字方式布置，外墙采用清水红砖。

Finished in November 1932, Saint Sophia Cathedral is located in Daoli District of Harbin. It is a Russian Orthodox cathedral in Neo-Byzantine style. The church is 53.35 meters high, covering an area of 721 square meters. Capped with a huge "onion-styled", the church has four buildings with yurt-shaped roofs of different size, which are orderly proportioned. Staircases are used to connect the four satellite buildings. The architectural plan is arranged as a Greek cross. The exterior walls are of red bricks.

立面装饰 / 穹顶细部

中央大学旧址
Site of the Central University

中央大学现为东南大学，保存至今的建筑有体育馆、图书馆、江南院、金陵院、中大院、大礼堂、南校门、生物馆、科学馆、梅庵等。其由杨廷宝、李宗侃等人设计， 于 1922—1933 年陆续建成，校园面积为 35.4 万平方米，建筑面积为 22 万多平方米。建筑均为西方古典形式，除体育馆为砖木结构外，其余均为钢筋混凝土结构。

The Central University was located at the place where is now Southeast University. The buildings preserved from the period of the Central University include the stadium, library, Jiangnan Court, Jinling Court, Zhongda Court, grand auditorium, south gate, Biography Building, Science Hall, Mei'an ("Plum Hut"),etc. They were designed by Yang Tingbao, Li Zongkan, etc. and completed in succession from 1922 to 1933. The campus covered an area of 354 thousand square meters, with a floor area of over 220 thousand square meters. All the buildings are in Western classical style. Except the stadium which is of masonry and timber structure, all are of reinforced concrete structure.

1933

上海市政府大楼落成。
南京中央体育场建成。
天津耀华学校建成。

图书馆旧影 / 大礼堂旧影

入口牌楼 / 观测点

紫金山天文台旧址

Site of Zijinshan Astronomical Observatory

紫金山天文台位于南京市玄武区紫金山上，被誉为“中国现代天文学的摇篮”。1929 年开始筹建，1934 年建成。天文台第一观象台是最早的建筑，其沿轴线对称布置，考虑到地势的高差，入口台阶处设置牌楼。建筑与周边环境融为一体，显得庄重朴实。

Zijinshan Astronomical Observatory, sitting on Zijin Mountain, Xuanwu District, Nanjing. It has been reputed as the "cradle of Chinese modern astronomy". It started to be planned in 1929 and was completed in 1934. The first observatory was the earliest building, which is laid out symmetrically along the axial line. Considering the difference in elevation, an archway was built at the entrance to the stairs. The building, which fits in well with the environment, appears dignified and simple.

1934

何立蒸（1912—2005）在《中国建筑》上发表《现代建筑概述》。
青岛圣弥爱尔教堂建成。
南京国民政府外交部大楼落成。
新华信托储蓄银行天津分行大楼建成。
浙江杭州钱塘江大桥动工兴建。此桥为中国近代史上第一座特大铁路公路两用桥。
雕塑家刘开渠创作的杭州第八十八师淞沪战役阵亡将士纪念碑建造完成，此为我国第一座表现抗日战争的纪念碑。
上海虹桥疗养院建成。

国际饭店

Park Hotel

国际饭店位于上海市南京西路，占地1300平方米，1934年由金城银行、盐业银行、中南银行、大陆银行等四家银行的储蓄会投资兴建，共24层，高82米，为钢架结构，是当时东亚最高的建筑。其底层外壁饰以黑色花岗石，上部全饰以褐色面砖。其室内外装修精致，几乎是美国20世纪20年代摩天大楼的翻版。

Located on Nanjing West Road, Shanghai, covering an area of 1,300 square meters, Park Hotel was built in 1934. With an 82-meter-high 24-storied structure of steel frame construction, it was then the tallest building in East Asia. The exterior walls are decorated with black granite at the bottom and paved with brown tiles above. The building features exquisite indoor furnishings. This is almost a copy of skyscrapers built in the 1920s in the United States.

入口立面

武汉大学早期建筑

Early Buildings of Wuhan University

1935

南京建成“国民大会堂”。
上海市图书馆、博物馆建成。
南京灵谷寺国民革命军阵亡将士公墓落成。
重庆美丰银行建成。

武汉大学早期建筑位于武汉市东湖湖畔。武汉大学早期建筑的设计思想代表了当时一流校园的理想模式。校园布局贯穿着中国传统建筑“轴线对称、主从有序、中央殿堂、四隅崇楼”的思想，采用“远取其势，近取其质”的手法，整体规划因山就势，散点布局，变化有序，整个建筑群就像一座座“花园”和“宫殿”。

The early buildings of Wuhan University abut Donghu Lake in Wuhan. The design idea reflects the ideal campus model of the best university at that time. The Chinese traditional architectural philosophy of “symmetry, priority, and main hall at the center with high buildings at the four corners” was adopted thoroughly in the design. In actual implementation, the design “making the best of the terrain in the distance and the material nearby”, meaning the overall layout pertaining to harmony with the circumstances. The group's feature is like a “garden” and a “palace”.

行政楼（原工学院）/ 图书馆 / 男生宿舍与图书馆 / 校园早期规划

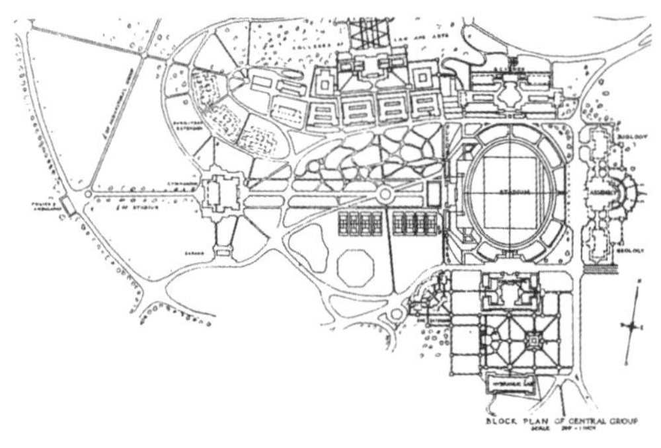

与外滩拍摄全景图

上海外滩建筑群

Architectural Complex of the Bund of Shanghai

有“万国建筑博览群”之称的外滩位于上海市黄浦江与苏州河的交汇处，它北起北苏州路，南至金陵东路，长约1800米。著名的中国银行大楼、和平饭店、江海关大楼、汇丰银行大楼再现了昔日“远东华尔街”的风采。这些建筑虽不是出自同一位设计师之手，也并非建于同一时期，然而建筑色调却基本统一。

The Bund, renowned as an expo of architecture from different nations, is located at the intersection between the Huangpu River and Suzhou River in Shanghai. It is an area stretching about 1,800 meters from North Suzhou Road in the north to Jinling East Road. Such famous buildings as the building of the Bank of China, Peace Hotel, Customs House and HSBC Building bear evidence of Shanghai's past as "Wall Street of the Fast East". Though these buildings were not designed by the same architect or built in the same period, their tones are basically uniform.

1936

广州《新建筑》创刊。

上海中国银行总行大楼建成。

上海法国邮船公司大楼建成。

南京中央博物院大楼完成设计，但迟至1947年完成建造。

长春伪满国务院、最高法院等由日本人设计建成。

建筑沿街立面细部构造及街景（组图）

南京西路建筑群
Architectural Complex on Nanjingxi Road

20 世纪 30 年代上海市跑马厅北侧的南京西路是当时上海市具有标志性的市中心。其主要建筑有国际饭店、大光明电影院、跑马厅大厦等。建筑呈现出多种风格、多种形式，既有外国建筑师的作品，也有中国建筑师的创作，这一组建筑实际上成了 20 世纪 30 年代中国和东亚娱乐消费文化的象征。

As a landmark of Shanghai downtown area in the 1930s, Nanjingxi Road, north to Race Club of that time, is the location of International Hotel, the Grand Theatre, Shanghai Race Club, etc. The architectural complex has embraced a variety of styles and forms, including works of foreign architects and Chinese ones, which are collectively a symbol representing Chinese and East Asian entertainment consumption culture in the 1930s.

金陵大学旧址

Site of the University of Nanking

金陵大学旧址位于江苏省南京市，现为南京大学鼓楼校区。建筑主要建于1916—1937年。建筑以中国传统建筑风格为基调，主要建筑沿一条南北向的主轴线布置，建筑物之间安排有形状规则的绿地、广场，与美国大学校园的特点相似。金陵大学具有代表性的建筑有北大楼、东大楼、图书馆等十余幢建筑。

The site of the University of Nanking, located at Nanjing City, Jiangsu Province, is now Gulou Campus of Nanjing University. The buildings of the University was built during 1916-1937. It adopts predominantly the Chinese traditional architectural style. The main buildings are laid out along a north-south main axis. Between them there are green spaces and squares of regular shape. In this respect the University of Nanking is similar to the campus of US universities. There are over ten buildings representative of the University of Nanking, i.e. the North Building, East Building and library.

1937

宁波镇海口海防设施建筑竣工。

中国营造学社梁思成、林徽因、莫宗江等在山西五台山发现唐代木构建筑佛光寺。

“七七”卢沟桥事变中宛平龙王庙、城南药王庙、卢沟桥南侧河神庙、西侧大王庙、北侧迴神庙以及宛平城内兴龙寺、观音庵、城隍庙、马神庙、九神庙等数十处古建筑毁于战火。

上海“八一三”抗战中，上海市大量近现代西式建筑、中西合璧式建筑和部分古建筑被毁或损伤严重。

云南数十万军民动工兴建滇缅公路。

重庆陆续将南岸建筑群等改建、扩建为政治、军事中心建筑群。

南京保卫战中，南京市三分之一的建筑物和财产化为灰烬；包括中山陵、夫子庙等在内的大量文化遗产被践踏、亵渎。

天津工商学院建筑系成立。

被绿色植被覆盖的北大楼

钱塘江大桥

Qiantang River Bridge

钱塘江大桥位于浙江省杭州市西湖之南，是第一座我国自行设计、建造的双层式铁路、公路两用桥。1937 年 9 月 26 日建成。大桥全长 1453 米。正桥有十六孔，桥墩十五座，雄伟壮观。

On south of West Lake in Hangzhou City of Zhejiang, the Qiantang River Bridge is the first double-deck railway-highway bridge designed and built independently by Chinese. It was completed on September 16, 1937. The bridge is 1,453 meters long.

钱塘江大桥（组图）

阳光落在室内阶梯上／建筑与外环境

1938

滇缅公路畹町至昆明段建成通车。至年底，全线贯通。
平西房山云居寺遭日军重型轰炸机轰炸，辽代建筑南塔被彻底毁灭。
河北阜平县普佑寺被日军彻底焚毁。
长沙岳麓山忠烈祠动工兴建。
中国营造学社刘敦桢、陈明达、莫宗江等考察云南昆明、大理、丽江等地古建筑遗存。
天津利华大楼建成。

南泉抗战旧址群

Nanquan Architectural Complex during Resistance War against Japan

南泉抗战旧址群包括林森别墅（听泉楼），孔祥熙官邸（孔园），校长官邸（小泉总统官邸），陈立夫、陈果夫官邸（竹林别墅），中央政治大学研究部（彭氏民居）等5处遗址。

Nanquan Architectural Complex during Resistance War against Japan consists of five sites of Linsen Villa (Tingquan Building), Kong Xiangxi's Official Residence (Kong's Garden), President's Official Residence (President's Xiaoquan Official Residence), Chen Lifu and Chen Guofu's Official Residence (Zhulin Villa), Political Research Department of the Central University (Peng's Residence).

厂区高炉旧影 / 厂区旧址（组图）

1939

南京国民政府教育部以“确定标准，提高程度”为目标，为全国大学制定并颁布的建筑专业课程统一教程，由梁思成、刘福泰、关颂声共同制定。
海口琼海关大楼落成。

1940

滇缅公路沿途屡遭日军空袭，多处桥梁被炸毁，均及时修复。
之江大学建筑系成立。
昆明南屏电影院建成。
天津望海楼教堂重建。

重庆抗战兵器工业旧址群

Chongqing Ordnance Industry Relics during Resistance War against Japan

重庆抗战兵器工业旧址群主要包括兵工署第一工厂抗战生产洞、兵工署第十工厂抗战生产洞等。这些遗址分布于重庆市江北区、沙坪坝区、九龙坡区、大渡口区和万盛经济开发区。现存的抗战生产洞数量超过 100 个，多数洞体保存较好。

Chongqing Ordnance Industry Relics during Resistance War against Japan include Ordnance Department's 1st Factory Production Cave, 10th Factory Production Cave, etc. These relics are scattered around Chongqing's Jiangbei District, Shapingba District, Jiulongpo District, Dadukou District, and Wansheng Economic Development Zone. The existing caves are over 100, most of which are well-preserved.

建筑中厅／外立面局部／入口立面

中国营造学社旧址

Site of the Society for the Study of Chinese Architecture

中国营造学社旧址位于四川省宜宾市李庄镇上坝村月亮田，建筑面积 349 平方米。中国营造学社的梁思成、刘敦桢、林徽因等中国建筑史学界的前辈在 1940 — 1946 年间都曾在此工作、学习和生活。

With a floor area of 349 square meters, the site of the Society for the Study of Chinese Architecture, at Yueliangtian of Shangba Village in Lizhuang Town of Yibin City, Sichuan Province. This is where the forerunners of Chinese modern architecture history such as Liang Sicheng, Liu Dunzhen, Lin Huiyin worked, studied and lived during 1940-1946.

同盟国中国战区统帅部参谋长官邸旧址

Official Residence of the Chief of Staff of the Allied Powers' Supreme Command in the China Theater

同盟国中国战区统帅部参谋长官邸旧址位于重庆市渝中区嘉陵新路63号，是抗战时期担任中国战区参谋长的史迪威将军的办公和居住地。该建筑为带地下防空洞的西式二层小楼，具有现代建筑之意。

At No. 63, Jialingxin Road, Chongqing, the Official Residence of the Chief of Staff of the Allied Powers' Supreme Command in the China Theater was General Joseph Warren Stilwell's office and residence. The building is a two-storied western-style villa with an underground air-raid shelter. The design emanates a modern taste.

入口立面 / 次入口及环境 / 原军事会议厅展室

1941

国民政府陪都建设计划委员会成立，开启了战争中的建设序幕。该委员会在此后三年内完成《陪都建设计划工作纲要》《战时建设计划大纲》《战时三年建设计划大纲》《陪都防空建筑改进意见》《测量陪都旧城区情况》等，并曾数次举办陪都建设设计展览。

经三次长沙会战，长沙城仅存中山亭等少数完整建筑物。

故宫及北平城市中轴线建筑测绘工作开始。

1942

林克明在《新建筑》杂志发表《国际新建筑会议十周年的纪念感言》，积极倡导现代建筑运动，批判"固有之形式"的官方倡导者和过度迎合业主的建筑师。

湖南武冈黄埔二分校中山堂落成。

延安中共中央大礼堂落成。

上海圣约翰大学工学院建筑系成立。

延安革命旧址

Historical Yan'an Revolution Base

延安革命旧址包括凤凰山中共中央旧址，杨家岭中共中央旧址，枣园中共中央书记处旧址，王家坪中共中央军事委员会、八路军总司令部旧址，陕甘宁边区政府旧址等。毛泽东等老一辈革命家在这里生活、战斗了13个春秋，这里见证了一系列影响和改变中国历史进程的重大事件。

The Historical Yan'an Revolution Base includes the site of the CPC Central Committee Office in Fenghuang Mountain, the site of the CPC Central Committee Office in Yangjialing Hill, the site of the CPC Central Committee Secretariat at Zaoyuan Garden, the site of the CPC Central Military Commission at Wangjiaping, the Eighth Route Army HQs, and the Shaanxi-Gansu-Ningxia Border Region Government. Mao Zedong and other famous revolutionists had lived and worked here for 13 years. Here witnessed a series of major events that swayed and changed Chinese history.

忠烈祠庭前甬道

1943

南岳忠烈祠落成。此建筑群是参战各国在二战期间兴建的规模最大的纪念建筑。

云南平彝县（今富源县）中山礼堂奠基。

四川民众自带口粮和工具，90天内修筑完成了成都等地的六座军用机场。

昆明西园别墅竣工。

1944

在“长衡会战”中，千年古城长沙历经四次战火，几成废墟；同为千年古城的衡阳，战后无一完整建筑。至此，湖南长沙、常德、衡阳三城与南京等成为“二战”中受损最严重的文化名城。

昆明中山纪念堂奠基。此建筑建成后被命名为抗战胜利堂。

梁思成编撰《中国建筑史》完稿（与莫宗江等合作）。

1945

日军在投降前夕炸毁731部队实验室等反人类罪证，并炸毁河北安平圣姑庙等大量珍贵文化遗产。

芷江机场中国军队指挥机构、南京中央军校大礼堂、北京紫禁城太和殿、广州市中山纪念堂、武汉中山公园、台北中山堂、天津法租界公议局、香港总督署等各类地方标志性建筑成为接受日军投降场所。

长春市苏军烈士纪念塔建成。

1946

《中国营造学社汇刊》第7卷第2期刊载林徽因专稿《现代住宅设计的参考》，系统介绍住宅社会学并反映时代潮流下的建筑思想。

清华大学建筑系成立。

昆明人民胜利堂落成。

南岳忠烈祠
Nanyue Martyrs' Shrine

南岳忠烈祠位于湖南省衡阳市南岳衡山香炉峰下方，是为纪念抗日阵亡的将士而建的，祠为中国现存有关抗战最大的纪念建筑群，是当年唯一以国民政府名义建造的抗战烈士纪念陵园，1943年6月全部竣工。其主体建筑风格和规制均仿南京中山陵的忠烈祠，中轴线沿山势而上，依次分布着拱门牌坊、七七纪念碑、纪念堂、安亭战役纪念亭和享堂等五个主体建筑。

At the foot of Mount Xianglu of Hengshan Mountain, the Martyrs' Shrine was built to honor and commemorate the martyrs of the Resistance War against Japan. It was the largest memorial architectural complex in existence, and the only one Martyrs Memorial Cemetery for the Resistance War against Japan built by the National Government of the Republic China, and was completed in June 1943. The main body is styled with a similar configuration as that of the Mr. Sun Yat-sen's Mausoleum in Nanjing. Along the central axis with the terrain sit five main architectural pieces, i.e. the decorated archway (paifang), Monument of July 7th Incident, memorial hall, Memorial Gazebo of Anting Battle, and the enshrining hall.

重庆人民解放纪念碑

Chongqing People's Liberation Monument

1941 年重庆市中区都邮街广场建成了一座碑形建筑，名为“精神堡垒”。抗日战争胜利后，重庆市决定在“精神堡垒”的旧址上建抗战胜利纪功碑，1947 年 8 月竣工，采用钢筋、水泥建造，高 27.5 米，为八角形柱体盔顶钢筋混凝土结构。1949 年 11 月 30 日重庆解放，抗战胜利纪功碑改名为人民解放纪念碑（简称解放碑）。

A monument-style structure was built on Duyou Street Square in the Central District, Chongqing City on 1941 and was named "Spiritual Fortress". After the victory of the War of Resistance against Japanese Aggression, Chongqing Municipal Government decided to build a credit monument. The construction was completed in August, 1947. The monument, built with reinforced concrete, is 27.5 meters tall with an octagonal helm-style top. On November 30th, 1949, Chongqing was liberated. The new inscription the People's Liberation Monument ("Liberation Monument" for short) replaces the former one after then.

1947

东北烈士纪念塔在哈尔滨市举行奠基典礼。

1948

上海都市计划委员会修编完成《上海都市计划总图草案报告书（二稿）》。

1949

华北人民政府文物处刊发梁思成编写的《全国建筑文物简目》。
北平市成立都市计划委员会。
人民英雄纪念碑奠基典礼举行。
中华人民共和国成立。
中央人民政府政务院及政务院财政经济委员会（时称“中财委”）成立。中财委中央财经计划局设有基本建设计划处，主管全国基本建设、城市建设和地质工作。
公营永茂建筑公司成立。

解放碑正立面

民族传统创新

Innovation in National Tradition

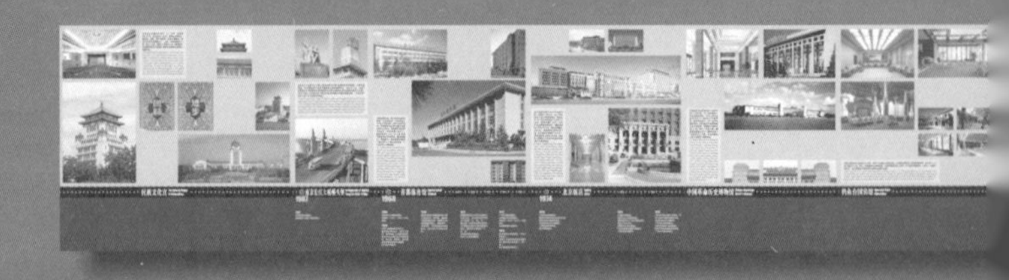

1950–1979

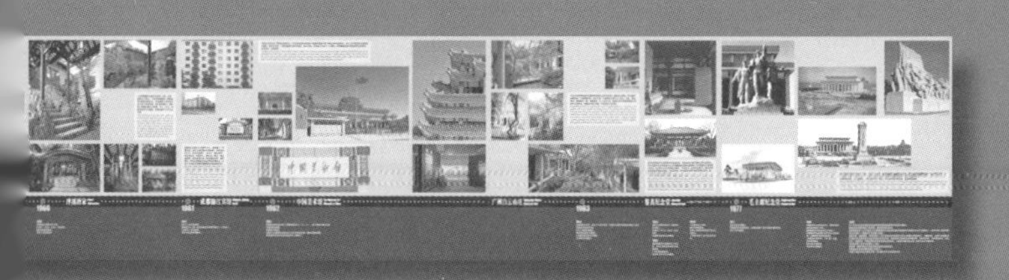

中华人民共和国成立后，受政治及经济环境的影响，以“社会主义内容、民族形式”为创作指针，国家提出了“适用、经济、在可能条件下注意美观”的建筑方针。中国建筑师在前辈开创的现代建筑风格的基础上，继续进行中国建筑风格的探寻与实践，不仅清除废墟、建立秩序，也坚守民族形式的主观追求。这一时期超越了历史上的任何时代，出现了许多令人瞩目的建筑，如人民大会堂、北京友谊宾馆、电报大楼、西安人民大厦、泮溪酒家、成都锦江宾馆。中国建筑师如张镈、戴念慈、林乐义、洪青、莫伯治、徐尚志等为之做出贡献。必须承认，受种种因素的影响，思想的固步自封和停滞对中国建筑发展的冲击很严重，中国建筑在艰难的前进中探索。

After the founding of People's Republic of China, under the influence of political and economic environment, the architectural principle of “Applicability, Economy, and Attention to Beauty Where Conditions Permit” was raised in accordance with the guideline of “Socialist in Content, National in Form” on creation. Chinese architects continued to carry out explorations and practices in Chinese architectural styles on the basis of modern architectural styles created by their predecessors. They removed the debris, established orders and adhered to the subjective pursuits of national forms. This period surpassed any times in history and witnessed the emergence of many remarkable buildings such as the Great Hall of the People, Beijing Friendship Hotel, the Telegraph Building, Xi'an People's Hotel, Panxi Restaurant and Chengdu Jinjiang Hotel. Chinese architects including Zhang Bo, Dai Nianci, Lin Leyi, Hong Qing, Mo Bozhi and Xu Shangzhi made contributions to them. It has to be admitted that affected by various factors, the restriction and retention of thoughts exerted serious impacts on the development of Chinese architecture, which explored the way forward amid difficulties.

设计效果图 / 外立面旧影

北京和平宾馆
Beijing Peace Hotel

北京和平宾馆位于北京市东城区金鱼胡同，其在设计之初为中等旅馆，后为满足在北京召开的“亚洲及太平洋地区和平会议”的使用需求，对原设计进行了部分修改。和平宾馆的设计周到、合理，空间组织紧凑，用不对称手法安排建筑的出入口。建筑立面简洁大方，与环境巧妙结合，造价经济，与当时国家的经济形势和政策十分吻合，是中国当代建筑的经典之作。

Beijing Peace Hotel, located in Jinyu Hutong, Dongcheng District, Beijing City, was initially designed as a medium-sized hotel. Later, to accommodate the Asian-Pacific Peace Conference, some modifications were made on the basis of the original design. Considerately and reasonably designed, Peace Hotel features a compact spatial plan. The entrances/exits are arranged in an asymmetrical fashion. The facade of the building looks simple and elegant and fits in well with the environment. It was a classic work of Chinese contemporary architecture.

1950

中共西南局决定建造重庆人民大礼堂、重庆西南局办公大楼和重庆市委会办公大楼等。邓小平为此提出雄伟壮丽与勤俭节约并重的建设方针。

中央人民政府文化部下设社会文化局（今国家文物局）。

陈植、赵深、童寯、蔡显裕等组建上海联合建筑师、工程师事务所，1952年停业。

西南建筑公司在重庆成立。

1951

梁思成、刘开渠、莫宗江等开始设计人民英雄纪念碑。

中华人民共和国第一座大型水库——官厅水库开工建设。

云南省建工局设计室（现云南省设计院）成立。

东北工业部颁发《基本建设工程设计暂行管理条例》。这是中华人民共和国成立后的第一个地方性设计管理条例。

上海新建的虹口体育场举行落成典礼。

大连人民文化俱乐部建成。

1952

中央人民政府建筑工程部成立。

第一次全国建筑工程会议召开。建筑工程部在第一次全国建筑工程会议后的一份报告中提出“……设计的方针必须注意适用、安全、经济的原则，并在国家经济条件许可的情况下，适当照顾建筑外形的美观，克服单求形式美观的错误观点”。

政务院财政经济委员会主任陈云正式发布命令，颁发《基本建设工作暂行办法》，并附“各种事业基本建设的限额”的规定。

北京和平宾馆建成。

住宅外观（组图）

曹杨新村
Caoyang New Village

该住宅区位于上海市西北郊，1953年竣工，是中华人民共和国第一个工人新村。一期占地23.63公顷，建筑面积为11万平方米，建成住宅4000套。其规划充分考虑用地的自然条件，自由布局，创造了一个环境优美宜人的住宅小区。住宅多为2～3层，平面简单适用，日照充足。

The residential community, located in the northwestern suburb of Shanghai, built in 1953. It was the first workers' new village in China. The first phase of the construction project, consisting of 4,000 apartments, covers an area of 23.63 hectares with a total floor area of 110,000 square meters. The design has given full consideration to the natural conditions. Adopting a free layout, created a residential community with a beautiful environment. The buildings, mostly 2-storied or 3-storied structures with simple and practical plans, enjoy plenty natural lighting.

西安人民大厦

Xi'an People's Hotel

西安人民大厦于1953年竣工，是当时我国著名的大型宾馆之一。人民大厦占地6.6公顷，拥有不同规格的客房600间，餐饮、娱乐、商务等配套设施齐全。建筑室内环境优雅舒适，室外庭院宽敞优美。建筑中部颇具雕塑感的突起令人耳目一新，建筑立面造型古朴、简洁。

As one of the large well-known hotels in China at that time, Xi'an People's Hotel was completed in 1953. The People's Hotel has 600 guest rooms of different grades, restaurants, entertainment and commercial facilities in its 6.6 hectares. The interior is elegant and comfortable. The courtyard is spacious and charming. The architect using a sculpture-like central section to deliver a novel appeal. The facade is pristine and simple, yet grand and magnificent.

1953

重庆西南局办公大楼和重庆市委会办公大楼竣工。

国家第一个五年计划（1953—1957）开始实施，其中包括苏联援建项目156项工程。政府决定将建设力量转向工业建设，此后，各大区先后建立的国营设计院多改为“工业建筑设计院”。

国家计委召开全国勘察设计计划会议，布置1954年度设计工作计划的编制工作。这是我国首次编制全国范围的设计计划。

北京友谊医院建成。

西安市委礼堂建成。

中国建筑学会第一次代表大会在北京文津街中国科学院院部正式开幕。周荣鑫在报告中提出“以适用、经济、美观为原则”（之后，建工部于1955年2月将建筑设计方针修订为“适用、经济，在可能的条件下注意美观”；2016年2月6日中共中央国务院在《进一步加强城市规划建设管理工作的若干意见》中再次提及并确定“适用、经济、绿色、美观”的新八字建筑方针），梁思成发表题为《建筑艺术中社会主义现实主义的问题》的报告。

华东建筑设计公司与南京工学院合办的中国建筑研究室成立。

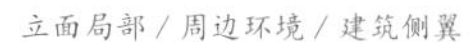

立面局部 / 周边环境 / 建筑侧翼

正立面及市民广场 / 鸟瞰

1954

在第一届全国人民代表大会第一次会议上，周恩来总理在《政府工作报告》中批评太原热电厂建设中惊人的浪费现象。
重庆人民大礼堂建成。
国家建设委员会成立。
《建筑》杂志创刊。
在《建筑学报》创刊号上，梁思成发表题为《中国建筑的特征》的论文。
天津市第二工人文化宫建成。
北京钢铁学院建成。
中直礼堂建成。
陕西省建筑工程局办公楼建成。
中国建筑东北设计研究院办公楼建成。

北京展览馆
Beijing Exhibition Center

北京展览馆是 20 世纪 50 年代北京建造的第一座大型公共建筑。其平面呈“山”字形，左右对称，前后呼应，前伸的两翼为正门入口的圆形广场，广场正中是一个直径为 45 米的巨大喷水池。主馆包括展览大厅、剧场、电影厅、餐厅、露天展览场等。

Beijing Exhibition Center is the first large-scale public building constructed in Beijing in 1950s. With a trident-shaped plan, the main construction is in axial symmetry. The two wings extend forward to form a round square at the main entrance. In the center of the square there is a gigantic fountain pool with a diameter of 45 meters. The main venue consists of the exhibition hall, the theatre, the cinema hall, the dining hall and the open-air exhibition area.

北京儿童医院
Beijing Children's Hospital

北京儿童医院为北京市最大的儿童专科医院，医疗区分为南、北、中三部分，门诊部居中，各科为独立单元，双重走廊两次候诊，并有家长候诊区。建筑外立面古朴中透着简洁的现代气息；烟筒水塔合二为一，并以方塔造型加以装饰，后被拆除。北京儿童医院是当年中国建筑师探索中国现代建筑的优秀实例。

As the largest children's hospital in Beijing, the medical service area can be divided into three sections. The clinic is at the center and each department has an independent unit. There are two aisles and a two-phase waiting process. There are also places for parents to wait. The courtyard of the hospital is very spacious and the buildings are flexibly laid out. The facade is simple and modern; the chimney and water tower was combined and decorated into one square tower structure. Beijing Children's Hospital is a fine example of Chinese architects' exploration in Chinese modern architecture.

儿童医院水塔 / 全景鸟瞰

同济大学文远楼

Wenyuan Building of Tongji University

文远楼位于同济大学校园内，建于 1954 年，为三层框架结构，最初作为同济大学建筑系馆使用。其平面布局按功能需要布置，建筑内交通路线简洁、通畅，室内踏步和楼梯扶手栏杆的处理干净、流畅，并加以简单装饰；建筑入口处理灵活，入口上方的雨棚造型新颖；细部处理得当、舒适，表现出了建筑师对现代中国建筑发展的探索和追求。

In a three-storied frame structure, originally used by the Department of Architecture, on the campus of Tongji University, the Wenyuan Building was built in 1954. The plan was arranged for function. The internal traffic arrangement is simple and efficient. The treads and rails of the staircases are simple and fluent with plain decorations. The entrance was handled exquisitely, with the canopy at the entrance of a novel design. All details are perfectly done, revealing the architects' pursuit and exploration of Chinese modern architecture.

文远楼主入口侧外观

主立面

西安人民剧院

Xi'an People's Theater

西安人民剧院于 1953 年开始建设，1954 年 8 月落成。建筑外观古朴、庄重。4 根大红圆柱立于入口中央，梁枋上的油漆彩绘表现出传统建筑文化元素，又有极强的装饰作用，赋予建筑以特殊的文化气质，成为古都西安的一大景观。剧院内部观众厅分为两层，视听效果颇佳。

The construction of Xi'an People's Theater began in 1953, and was completed in August 1954. The building is quaint and solemn. Four crimson-colored cylindrical columns in the center of the entrance are accentuated with strong traditional architectural elements that are lacquer color-painted lintels, beams and architrave, which together give an unusual cultural disposition to the theater. It has become a scenic spot in the ancient capital, Xi'an. The design of the interior two-storied auditorium produces excellent visual and acoustic effect.

全景鸟瞰 / 友谊宫旧影 / 主楼外立面旧影

北京友谊宾馆
Beijing Friendship Hotel

北京友谊宾馆总占地面积33.5万平方米，是一座全对称、具有浓郁的古典风格、富有气势的绿色琉璃瓦大屋顶式建筑，雄伟壮观。整个建筑群突出了中国的民族传统，围合出一个四合院，南北“工”字形配楼呈严格对称的扇形布局，轴线两侧五座民族色彩浓厚的大楼均为绿色琉璃屋顶，飞檐流脊、雕梁画栋。

Covering a total area of 335,000 square meters, Beijing Friendship Hotel is a magnificent building of a distinctive classic style in absolute symmetry, featuring an imperial roof paved with green-glazed tiles. The entire architectural complex highlights Chinese national tradition by enclosing a quadrangle. With the VIP Building and Friendship Palace sitting on the middle axis, the complex, including the I-shaped wings in south and north , has a strictly symmetrical fan-shaped plan. The five characteristically national-style buildings along the two sides of the axis are all paved with green-glazed roofs, featuring “flying eaves” and “flowing ridges”, “carved beams” and “painted rafters”.

重庆市人民大礼堂

The People's Great Hall of Chongqing

重庆市人民大礼堂是中华人民共和国初期的建筑代表作之一。其设计仿明、清坛庙建筑形式，采用轴向对称的传统手法，结构匀称、布局严谨。主体部分的穹庐金顶，脱胎于北京天坛的祈年殿，由大礼堂和东、南、北楼四大部分组成。其总占地面积为6.6万平方米，其中礼堂占地1.85万平方米。礼堂高65米，建成后进行过多次维修和改造。

The People's Great Hall of Chongqing is an imitation of the palaces of the Ming and Qing dynasties. With a layout in axial symmetry, it is well-proportioned and meticulously planned. The golden-topped dome is modelled after Qinian Hall at the Temple of Heaven. It consists of the auditorium, the east, south and north buildings. The complex covers an area of 66,000 square meters, with the auditorium covering an area of 18.5 thousand square meters. The auditorium is 65 meters high. After completed, the building has been through multiple repairs and transformations.

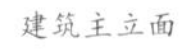
建筑主立面

外立面及挑檐局部 / 入口局部 / 入口立面

天津大学主楼

Main Building of Tianjin University

天津大学主楼即天津大学第 9 教学楼，坐落于校园教学区的中部。建筑正面南临广场，环境优美，严整亲切。建筑以民族形式为创作基础，突出的坡屋顶中部采用了十字交脊的歇山造型，在传统做法上有所创新。建筑外立面为三段式结构，造型稳重，比例匀称，端庄、典雅、庄重、平和，既有中国传统文化的高贵品质，又有一定的时代气息。

Main Building of Tianjin University is the teaching building No. 9 of Tianjin University, which is located in the central area of the campus. The front facade faces the square to the south. The campus is beautiful, orderly and friendly. Styled in Chinese traditional form, the pitched roof has a cross-ridged hip-and-gable section, an innovation in the otherwise pure traditional building. The facade is three-sectioned and composed with good proportion, emitting a demure, graceful, solemn, and peaceful sensation, an infusion of the noble temperament of Chinese traditional culture and modern times.

观礼台沿河侧外观（组图）

1954

天安门观礼台

Tian'anmen Reviewing Stands

天安门观礼台位于天安门前方东西两侧，主要用于国庆等重大庆典观礼。观礼台东西对称，各有7个台。天安门观礼台起初为举行开国大典时临时搭起的砖木结构建筑，1954年，在原有的基础上改建为砖混结构的永久性观礼台。观礼台呈北高南低的倾斜式，内有梯形台阶，总容量为21000人。

Tiananmen Reviewing Stands are on the east and the west sides of Tian'anmen Tower for major occasions like the National Day celebration. The seven stands on each side are symmetrical. Originally, they were temporarily built with brick-and-wood materials. In 1954, permanent stands of brick and concrete were built on the original foundation.All stands are high in the northern side and low in the southern side, with staircases inside and a capacity of 21,000 people.

宿舍屋顶及挑檐 /3 号宿舍楼外观

清华大学 1~4 号宿舍楼

Tsinghua University's Dormitory 1~4

清华大学 1~4 号宿舍楼在整体规划上采用了轴线对称的传统布局，建筑间有围合而成的庭院。建筑立面的处理手法借鉴了西方古典主义建筑传统，在垂直方向上采取三段式设计，在水平方向上利用轴线中心线的布置突出整组建筑严谨的格局。

In a traditional axial-symmetry, Tsinghua University's Dormitory 1~4 are of courtyard configuration surrounded by buildings. The facades are treated in Western classicism architecture style. The facade is divided into three sections vertically. The arrangement of using central line horizontally highlights the precise structure of the buildings.

1937 年前的群贤楼及周边建筑 / 建南大会堂

1954

厦门大学早期建筑
Early Buildings of Xiamen University

厦门大学本部位于厦门岛东南端，占地 2500 多亩（1 亩 =667 平方米），被誉为中国最美的大学之一。学校始建于 1921 年，主要建筑包括群贤楼群、建南楼群、芙蓉楼群等。厦门大学的建筑注重闽南式大屋顶与西式建筑特点的巧妙结合，富有闽南地域风格，又结合了西方建筑的式样，形成了中西混合的独特建筑形式。

At the southeastern corner of Xiamen Island, covering an area of 2,500 mu (1 mu = 667 m^2), Xiamen University is praised as one of the most beautiful college campuses in China. Built in 1921, the main construction consists of the Qunxian, Jiannan, and Furong complexes. With the big roof characteristic of southern Fujian and western architectural features, the buildings of Xiamen University are in local style with western architectural forms. The final architectural complex is a product of a unique version of an integrated Chinese-Western architectural genre.

集美学村
Jimei School Village

集美学村位于厦门集美半岛，坐落于集美村，由著名爱国华侨陈嘉庚先生于1913年始倾资创办，享誉海内外。之后几十年陆续建成各类学校，形成了由学前教育至小学、初中、高中，从本科教育到硕士、博士教育的人才培养体系。其中集美中学南薰楼高15层，是集美学村的制高点。

Jimei School Village is in the Jimei Village on Xiamen's Jimei Peninsula. It was built by the famous patriotic overseas Chinese Chen Jiageng in 1913. The school village is well-known in China and abroad. During the following decades, many schools were built in succession. Forming an educational system from elementary schools to high schools, from undergraduate education to postgraduate education. Among the buildings, Nanxun Building of the Jimei Middle School is 15-storied, which is the highest point of the school village.

道南楼

剧场及周边建筑环境／建筑模型剖面

1955

首都剧场
Capital Theatre

首都剧场是中华人民共和国成立初期我国新建的首座以演出话剧为主的专业性剧场，各项设施均处于当时国内的先进水平，设有电动活动台门和电动太幕，并首次在舞台中央设置电动旋转台。剧场坐东朝西，从入口依次由前厅、观众厅、舞台和后台四部分组成。设计者在探索将中国传统建筑艺术手法应用于现代建筑中做出了很大努力。

Beijing Capital Theatre was the first specialized theatre hosting primarily drama performances built in the early years of the People's Republic of China. When built, it was among the leading theatres in every aspect, with an electrically-powered proscenium, an electrically-powered screen and an electrically-powered rotational platform at the center of the stage. The theatre faces west and consists of, from entrance to back, the front hall, audience hall, stage and backstage. The architects made tremendous effort to apply the techniques of Chinese traditional architecture to modern architecture.

1955

建筑工程部召开设计和施工工作会议，对建筑中的形式主义、复古主义和浪费现象展开了严肃、全面的批评。
建筑工程部办公大楼建成。
中共北京市委办公楼建成。
全国政协礼堂建成。
重庆市体育馆建成。
中央军委办公厅机关宿舍建成。

入口立面 / 外立面局部一 / 外立面局部二 / 入口雕花装饰

上海展览中心

Shanghai Exhibition Center

上海展览中心原为中苏友好大厦，1954 年 5 月 4 日动工兴建，占地 2.5 万平方米。其设计受苏联同类建筑影响很大，属典型的俄罗斯巴洛克建筑风格。建筑沿中轴线对称布局，并围合出三个主要的室外广场，建筑之间的连廊分隔出几个绿化庭院。一些细部加入了反映当时政治氛围的装饰符号、中国传统装饰和图案。

Formerly the Sino-Soviet Friendship Building, Shanghai Exhibition Center started to be built on May 4, 1954, covering an area of 25,000 square meters. The design shows great influence of Soviet buildings of the same type and in a typical Russian Baroque style. With a plan in axial symmetry, the buildings enclose three major outdoor squares. The connecting corridors are used to partition several courtyards with vegetation. The decoration details reflecting the contemporary political atmosphere and traditional Chinese decorative patterns.

北京市百货大楼
Beijing Department Store

北京市百货大楼坐落于北京市中心王府井，是 1949 年以后国内最早兴建的大型百货商场，1972 年又建设了附楼。主楼地下 1 层，地上 4 层，局部 6 层，附楼 5 层。建筑为矩形，中部高处设计为三开间空廊，以突出建筑的重点。檐口采用传统式样的额枋、雀替形式，局部饰以中国建筑纹样。

Beijing Department Store is situated at Wangfujing, downtown Beijing. It was one of the earliest large-scale department stores of China built after 1949. In 1972, the attached building was built. There is one story underground, four stories above ground and six stories in some parts of the main building; the attached building is five-storied. Rectangular in plan, the raised middle part is a three-bay "empty corridor" designed to give prominence. At the cornices, traditional architraves and decorated brackets (que-ti) are adopted and decorated with Chinese architectural patterns.

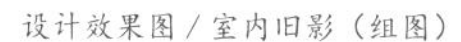

设计效果图 / 室内旧影（组图）

建设部办公楼

Office Building of the Ministry of Construction

建设部办公楼位于北京市百万庄，建筑风格融汇了西洋和中国传统语言。主楼高7层，南北配楼为5层，主体建筑面宽220米。立面为三段式，设有高台阶、大门廊，顶部的双层檐口加之椽子、瓦当、梁枋等民族传统式样体现了富有民族特色的造型。底层半圆拱窗套表现出雄浑稳固的特点，标准层窗下墙面上的简洁纹饰为建筑增添了精致的效果。

The Office Building of the Ministry of Construction is located at Baiwanzhuang, Beijing. Its architectural style is a combination of Western and Chinese traditional architectural elements. The main construction is seven-storied, whereas the south and north wings are five-storied. The main construction is 220 meters wide. The facade is three-sectioned. There are tall steps and a large porch. Traditional national architectural features including double eaves, purlins, decorative tiles and girders are employed, showing a nationally characteristic style. The semi-circular arched window pockets add to the sense of robustness of the building, whereas the simple patterns on the walls below the standard floor lend more exquisiteness.

立面装饰 / 檐口局部 / 入口立面

建筑远观旧影 / 西立面改造后 / 西立面旧影

北京“四部一会”办公楼

Office Building for Four Ministries and One Commission in Beijing

北京“四部一会”办公楼位于北京市复兴门外三里河。主楼地上6层，中部9层，是当时国内最高的砖混结构建筑。总平面布局采用当时流行的周边式，平面布置的特点是大进深。为使建筑有鲜明的民族形式的轮廓，各楼的主要入口部分加以双重檐庑殿攒尖顶屋顶，屋顶的承托部分自下而上收分，以衬得屋顶雄浑、壮观。

Located at Sanlihe, outside Fuxing Gate in Beijing. The main structure is six-storied above ground with a nine-storied middle section. It was then the highest building of brick and concrete in China. Thc overall plan adopts the then prevalent peripheral style, with the feature that the depth is deep. To show a distinctively Chinese national-styled contour, the main entrances to the buildings each has a double-eaved hipped roof with a pointed top. The support part of the roof tapers from bottom to top to set off the magnificence of the roof.

鲁迅先生之墓 / 鲁迅雕像

上海鲁迅纪念馆

Shanghai Lu Xun Memorial Hall

上海鲁迅纪念馆原与山阴路上海鲁迅故居毗邻，1956 年 9 月迁入虹口公园（今鲁迅公园）。上海鲁迅纪念馆集鲁迅生平陈列、鲁迅墓、鲁迅故居于一体。其规划设计采用中国造园艺术设计方法，布局自由流畅，注重交通路线，扩大原有水面，满足活动需求。

Originally abutting Lu Xun's former residence, the hall was relocated to Hongkou Pack (Lu Xun Park now) in September 1956. Lu Xun Memorial Hall exhibits Lu Xun's life events, Lu Xun's tomb, and Lu Xun's former residence. The plan and design are in Chinese garden style with emphasis on the unrestrained and fluent layout, traffic, water surface expansion, and activity needs.

1956

国家建委召开全国第一次基本建设会议。
中央党校办公楼建成。
中国合作总社办公楼建成。
杭州植物园建成。
上海虹口公园鲁迅纪念馆建成。
刘敦桢《中国住宅概说》出版。
国家建委颁发《勘察设计工作统一价目表》。
中山公园音乐堂建成。
天津市人民体育馆建成。

长春第一汽车制造厂早期建筑

Early Buildings of First Automobile Works in Changchun

长春第一汽车制造厂建于1953—1956年，长春第一汽车制造厂总共投资6.17亿元，建成生产区和生活区702480万平方米。长春第一汽车制造厂厂房呈矩形，为钢筋混凝土结构，屋架、天窗皆为钢制，厂房外立面为清水红砖，饰有柱廊，局部装饰有花饰、红旗、麦穗、齿轮等图案。

First Automobile Works (FAW) was built in 1953-1956, FAW invested 617 million yuan for her 7,024.8 square kilometers factory facilities and living quarters. The factory is a rectangular ferroconcrete structure with steel frames and skylights. The facade is made of red bricks decorated with pilasters partly patterned with flowers, red flags, wheat ears, and gears.

全景鸟瞰

观象台／建筑外阶梯／旧馆主入口

北京天文馆及改建工程

Beijing Planetarium and Renovation Project

北京天文馆位于北京西直门外大街，是国家级自然科学类专题科学博物馆。天文馆分天象厅、讲演厅、展览厅三个部分，南侧为有600个座席的天象厅，东、西方向各有一个展览厅。天象厅半球形的屋顶直径达23.5米，是天文馆的建筑特征。整个建筑造型简洁，装饰精美，比例和谐。天象厅屋顶采用的钢网结构由原德意志民主共和国（东德）帮助设计。

Beijing Planetarium, located on Xizhimenwai Street, Beijing, is a specialized national museum of natural science. The planetarium consists of three parts, i.e. the Astrology Hall, Lecture Hall and Exhibition Hall. On the south is the 600-seat Astrology Hall. There are two exhibition halls, one in the east and the other in the west. The semi-spherical roof of the Astrology Hall has a diameter of 23.5 meters—this is the most distinctive feature of the planetarium. The entire building is simple in form, exquisitely decorated and harmoniously proportioned. The planetarium roof adopts the steel grid structure. It was designed with the help of the German Democratic Republic (East Germany).

武汉长江大桥
Wuhan Yangtze River Bridge

武汉长江大桥，是中华人民共和国成立后在长江上修建的第一座复线铁路、公路两用桥。1957 年 10 月 15 日正式通车，全长 1670 余米。其上层为公路桥，下层为双线铁路桥，同时，大桥连接起中国南北的“大动脉”，对促进南北经济的发展、国民经济的建设起到了重要作用。

The Wuhan Yangtze River Bridge is the first double-deck two-line railway and highway bridge over the Yangtze River, built after the founding of China. The Wuhan Yangtze River Bridge put into use on October 15, 1957. The bridge is over 1,670 meters long. The upper deck is the highway bridge, and the lower deck is the two-line railway.

1957

武汉长江大桥建成。
北京天文馆建成。
中国建筑学会召开“中国建筑座谈会”，周荣鑫、梁思成、刘敦桢、龙庆忠、刘致平、陈明达、莫宗江、徐中、卢绳、汪之力等及苏联专家穆欣参加，并筹划编写《中国建筑通史》。
西安半坡遗址博物馆建成。

桥头堡室内局部 / 铁路桥 / 桥头堡侧立面 / 大桥远观

人民英雄纪念碑

Monument to the People's Heroes

人民英雄纪念碑位于北京天安门广场中心，分台座、须弥座和碑身三部分。台座分两层，四周环绕着汉白玉栏杆。下层须弥座束腰部四面镶嵌着10幅巨大的浮雕，生动而概括地表现了中国人民100多年来反帝反封建的伟大革命斗争史实。纪念碑正面（北面）镌刻着毛泽东同志题写的“人民英雄永垂不朽”八个鎏金大字，背面为毛泽东同志起草、周恩来同志书写的碑文。

Standing at the center of Tian'anmen Square in Beijing, the monument consists of three parts, i.e. the pedestal, sumeru base and body. The pedestal is two-layered and surrounded by white marble balustrades. The sumeru base is decorated on the four sides with ten white marble reliefs representing the great revolutionary struggles against imperialism and feudalism waged by the Chinese people in the past over 100 years. The front (north) is carried with the gilded eight-character inscription by Chairman Mao Zedong, which goes, "Eternal glory to the people's heroes!"; the back bears the epitaph composed by Mao Zedong and written by Zhou Enlai.

纪念碑旧影／纪念碑底座浮雕（组图）

北京自然博物馆
Beijing Museum of Natural History

1958

上海市建设委员会批准《上海市统一建造住宅暂行办法》。
哈尔滨市防洪纪念塔建成。
北京妇产医院建成。
中央广播大厦建成。
中国伊斯兰教经学院建成。

北京自然博物馆是中国依靠自己的力量建设的第一座大型自然科学类博物馆。建筑坐落于天桥南大街东侧，南北长 112 米，东西宽 61.2 米。整个建筑由两个展览楼和一个标本楼组成，其中主楼为两层，高 15.5 米，局部三层，高 24 米。建筑师将中国传统建筑元素与苏联建筑风格进行了比较完美的结合。

As the first large museum of natural science, Beijing Museum of Natural History was built independently of China. Erected by the east side of South Tianqiao Street, the building, with a length of 112 meters from south to north and a width of 61.2 meters from west to east, consists of two exhibition buildings and a specimen building. The main building is two-storied, with a height of 15.5 meters, and with some parts of the building is three-storied with a height of 24 meters. The architect did a seamless integration of the Soviet Union style and Chinese architectural elements.

入口外景 / 入口中厅 / 入口立面局部

沿街立面

北京电报大楼
The Telegraph Building, Beijing

北京电报大楼平面呈“山”字形，由电报机房和营业厅两部分组成，能够承担电报、长途电话和邮政等营业功能，还有召开全国电话会议的专用会议室。建筑主体和钟塔体形力求简洁优美，富有现代主义意味，同时细腻的色彩对比和比例推敲也符合中国的审美趣味。钟塔上装有直径 5 米的标准钟四面，整点报时或播放音乐，音响半径可达 2 千米。

With a trident-shaped elevation plan, the Telegraph Building consists of two parts, i.e. the telegraphic instrument room and the business hall, which provides services covering telegraphy, long-distance calls and post. Great effort was made to make the main body and clock tower of the building simple and beautiful and in a modern style. In the meantime, the contrast of fine colors and the proportion also conform to Chinese aesthetics. The clock tower is installed with four standard clocks with a diameter of 5 meters, which chime or play music on time. The sound can be heard 2 kilometers away.

大宴会厅一／大宴会厅二／北大厅阶梯／东立面局部

人民大会堂

The Great Hall of the People

人民大会堂位于北京天安门广场西侧，占地 15 公顷，平面呈“山”字形，南北长 336 米，东西宽 174 米，由万人大会堂、宴会厅、全国人大常委会办公楼三部分组成。大会堂平面对称，高低结合，台基、柱廊、屋檐采用中国传统的建筑风格。人民大会堂的建筑艺术造型和建筑材料与天安门城楼和广场协调一致而又有创新的效果。

The Great Hall of the People located on the western side of Tian'anmen Square in Beijing. Covering an area of 15 hectares, it has a trident-shaped plan which runs 336 meters from north to south and 174 meters from east to west. It consists of three parts, i.e. the Ten-thousand-People Great Hall, the Banquet Hall and the office building of the NPC Standing Committee. The Great Hall of the People features a symmetrical plan and combines high and low constructions. The terrace, colonnade and eaves are with Chinese traditional style. In terms of architectural form and materials, the Great Hall of the People agrees with Tian'anmen Tower and Square in style but with an effect of innovation.

1959

北京建成国庆十周年“十大建筑”：人民大会堂、中国革命历史博物馆、中国人民革命军事博物馆、全国农业展览馆、钓鱼台国宾馆、北京火车站、北京工人体育场、民族文化宫、民族饭店、华侨饭店。

刘敦桢等主编的《建筑十年》出版。

山西芮城永乐宫迁建工程启动。

中国建筑学会和建筑工程部在上海召开“住宅建设标准及建筑艺术座谈会”，刘秀峰部长作了《关于创造中国的社会主义建筑新风格》的报告。

天津宾馆建成。

武汉歌剧院建成。

上海锦江小礼堂建成。

北京火车站

Beijing Railway Station

北京火车站是中华人民共和国成立十周年北京“十大建筑”之一。中央大厅采用了预应力双曲扁壳屋盖，与立面中心的三大琉璃拱窗及其左右的两座琉璃钟楼有机结合，同时塔楼与钟楼以琉璃女儿墙连接起来，成为一个整体。整个立面主次分明，而且呈现出我国民族风格与现代技术相结合、新颖协调的艺术效果。

Beijing Railway Station was one of the ten greatest constructions built for the 10th anniversary of the founding of the People's Republic of China. The central hall features a pre-tension hyperbolic flat-shell roof, which fits well with the three large glazed-roof arched windows in the facades and the glazed-roof clock towers on the left and right; moreover, the towers and clock towers are joined with parapets to form an integral whole. Such a design gives prominence to the main construction and presents a harmonious novel artistic effect combined with Chinese national style and modern technology.

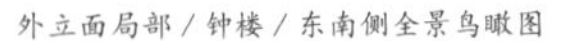

外立面局部 / 钟楼 / 东南侧全景鸟瞰图

远眺建筑上部 / 南立面局部装饰

民族文化宫
The Cultural Palace of Nationalities

民族文化宫建筑平面呈“山”字形，东西宽 185.78 米，南北进深 105 米。建筑由博物馆、图书馆、剧场、餐厅等组成，还有少量客房。民族文化宫东西两翼为 2 ~ 3 层，中部塔楼地下 2 层，地上 13 层，地面以上最高 67 米，挺拔高耸。全部墙面饰以白色面砖，屋顶采用翠绿色琉璃瓦，造型优美。

The Cultural Palace of Nationalities is 185.78 meters wide from east to west and 105 meters long from north to south. It consists of the museum, library, theatre, dining hall and a few guest rooms.The east and west wings of the Cultural Palace of Nationalities are 2 or 3-storied constructions. The tower at the middle has 2 stories underground and 13 stories aboveground. It is 67 meters tall at the highest point. All the walls are decorated with white tiles and the roofs are paved with green-glazed tiles.

全景旧影 / 西大厅

中国革命历史博物馆

Chinese Revolution History Museum

中国革命历史博物馆位于北京天安门广场东侧，与广场西侧的人民大会堂相对应。为与人民大会堂和天安门广场的巨大尺度相配，建筑师采用了“目”字形的建筑布局，利用建筑内院获得较大的外形轮廓。博物馆中央靠天安门广场一侧为一柱式门廊，建筑师希望通过这个门廊将天安门广场空间引入内院，并与对面的人民大会堂的门廊相呼应。

The Chinese Revolution History Museum located on the east of Tian'anmen Square in Beijing, it sits right opposite to the Great Hall of the People. To match the gigantic dimensions of the Great Hall of the People and Tian'anmen Square, the architects adopted a layout resembling the Chinese character “mu” so that the inner yards can be utilized to increase the profile of the building. In the middle of the building is a portico with columns, with which the architects hoped to merge the space of Tian'anmen Square with the inner yard and correspond with the portico of the Great Hall of the People.

钓鱼台国宾馆

Diaoyutai State Guesthouse

钓鱼台国宾馆占地面积42公顷，建有15栋独立的贵宾楼、2栋服务楼和警卫楼等附属设施，还开挖了人工湖，建造了小桥、亭台。各栋建筑以砖墙承重，为现浇钢筋混凝土结构。外立面采用釉面砖、花岗石、水刷石、剁斧石，室内地面多为硬木拼花地板或水磨石地砖。

Covering an area of 42 hectares, Diaoyutai State Guesthouse consists of 15 detached guesthouses, 2 service buildings and a security building, etc. Besides, there are also artificial lakes, bridges and pavilions. All the buildings, supported by brick walls, are of cast-in-place reinforced concrete. All the guesthouses are decorated with high-class materials then, the facade is of glazed tiles, granite, granitic plaster and chop axstone. As for interior decoration, the floor is paved with hard-wood parquet or terrazzo tiles.

室内环境（组图）/ 入口装饰 / 环境布置

全景鸟瞰图 / 餐厅空间组织 / 观众席

北京工人体育场
Beijing Workers' Stadium

北京工人体育场位于北京市朝阳区三里屯，是一座椭圆形建筑，设有观众席 8 万个。竞赛场设有田径比赛场地和足球比赛场地，南侧看台上设有大显示器和时钟。项目还配置了 1500 床的运动员招待所，建有室外训练场和游泳场。

Beijing Workers' Stadium, located at Sanlitun, Chaoyang District, Beijing City, is an oval structure building. As designed, there are 80,000 seats. There are grounds for track & field events and football. On the southern stand there is a large screen and a clock. There is also a 1,500-bed hotel for athletes. There is also an outdoor training and a swimming pool.

建筑夜景／内庭院夜景／内庭院阶梯

1960

泮溪酒家

Panxi Restaurant

1960

河南林县红旗渠动工兴建。

关颂声（1892—1960）先生逝世。

西安钟楼邮局建成。

北京二七剧场建成。

广州泮溪酒家全园分为厨房、厅堂、山池、别院四个部分，之间以游廊相连。顾客流线与输送流线分开，将交通的干扰和交叉减少到了最低。室内装修精巧，其中窗心隔扇的套色花玻璃、斗心、钉凸、木刻花罩等尤为精彩，有很高的手工艺艺术价值。

Panxi Restaurant has its inner space partitioned into the kitchen, dinner hall, interior landscape, and wing courtyard. The guest traffic is separated from the service traffic, so the interference from each other is reduced to a minimum. The overall layout is versatile yet integrated. The interior decoration is exquisite. The patterned stained glass and flower-pattern wood panels on the sash of windows are particularly attractive with high artifact value.

成都锦江宾馆

Chengdu Jinjiang Hotel

1961

北京工人体育馆建成。
国务院公布《第一批全国重点文物保护单位》（108处）。
上海闵行一条街建成。
清华大学主教学楼建成。

1961

成都锦江宾馆位于成都市中心，南临锦江，环境优美，设有总统套房及各类套房、标准客房500间，还有中西餐厅、会议厅、保龄球场、室内游泳池、购物中心、健身房、舞厅、咖啡茶座等，是西南地区第一座五星级宾馆。建筑体形、风格与周围的自然环境很好地结合，反映了20世纪中后期中国建筑创作的基本特色。

Facing the Jinjiang River to the south, surrounded by a picturesque environment, Chengdu Jinjiang Hotel is at the center of Chengdu. It has presidential suites and other kinds of suites, 500 standard guest rooms, Chinese and western dining rooms and restaurants, conference rooms, bowling alleys, indoor swimming pool, shopping center, gym, ballroom, and coffee shops. It is the first five-star hotel in southwest China. Revealing the basic features of Chinese architecture in the mid and late 20th century.

宾馆外观 / 宾馆门前标识

中国美术馆

The National Art Museum of China

1962

中国建筑研究室出版《中国建筑简史》（上、下），其下卷首次系统述论中国近现代建筑。

中国美术馆建成。

北京市土木建筑学会在劳动人民文化宫举办第一届建筑绘画展览。

华东地区住宅设计经验交流会在上海举行。

中国美术馆采用了典雅的民族形式，中间凸出的四层吸收了敦煌莫高窟九层飞檐的传统造型语言，设计为古典楼阁式的重檐大屋顶，其余为平顶。门廊及廊榭采用中式屋顶，略作点缀，在风格上形成了一个整体。外墙素陶面砖与琉璃面砖间杂使用，色彩和谐、壮观典雅。

The National Art Museum of China adopts the elegant national form. A four-layered projecting part in the middle, absorbing traditional features from the nine-storied tower with "flying" eaves of the Mogao Caves, features a double-eaved roof in the style of classical Chinese pavilion; the rest parts are with flat roofs. The porch and veranda are covered with traditional Chinese-style roof with simple decoration, which form a whole in style. The exterior walls are paved with plain-pottery tiles and glazed tiles in a mixed fashion—the colors are harmonious, spectacular and elegant.

美术馆主立面外观 / 主立面檐口局部

建筑围合庭院（组图）

广州白云山庄

Guangzhou Baiyun Mountain Villa

1963

辽宁工业展览馆建成。
西安电报电话大楼建成。
上海蕃瓜弄住宅小区建成。
杭州西泠饭店建成。

白云山庄是岭南园林与建筑完美结合的代表作品，建筑依山就势，空谷藏轩，回廊周布，空间错落有致，层次分明。三叠泉被从室外引入室内，水形、泉声、砖色、绿意融为一体，相辅相成，令人过目不忘。建筑体量与风格遵循现代功能主义的轨迹，着重表达现代功能、材料和技术的内涵。

Baiyun Mountain Villa is a representation work of Lingnan gardens and architecture. The buildings ride on the mountain with pavilions deep in the mountains and winding corridors in the peripheral. The spatial composition has layered partition. The water of a spring of a three-staircase waterfall is directed indoors. The flow of water, purl of stream, color of bricks, and prosperity of plants are infused together in a harmonious fashion, creating a lingering, mesmerizing sensation. The architectural volume and style reveal traces of modern functionalism that emphasizes connotations of modern functions, materials and techniques.

纪念堂内景 / 正立面

鉴真纪念堂

Jianzhen Memorial Hall

鉴真纪念堂参照日本招提寺的金堂设计，而金堂是以唐代佛殿为蓝本建造的，因此可以说鉴真纪念堂完美表现出了中国唐代木构建筑之风采。纪念堂为五开间庑殿顶建筑，气势宏大，室内色彩简洁，室内明间设有鉴真和尚楠木坐像，下有须弥座。纪念堂用步廊与前面的碑亭连在一起，形成庭院，使环境更显幽静。

The Jianzhen Memorial Hall was modeled after the Golden Hall of Japan's Toshodai-ji which was constructed based on Buddhist temples of the Tang Dynasty, we can say that it perfectly desplayed the style of Chinese wooden architecture in Tang Dynasty. The hall is a five-bay building with a hipped roof, emanating grandeur. It has coffers and an interior with simple coloring. The internal of the main bay has a sitting joss of Monk Jianzhen on a sumeru base. A roof corridor connects the hall and the front stele gazebo to form a courtyard configuration. The ambiance emits solemnity.

1964

清华大学建筑系创办《建筑史论文集》。
朱启钤（1872—1964）先生逝世。
中国民航总局办公楼建成。

1965

国家建委召开全国设计工作会议，对设计革命运动进行总结和研究。
广州友谊剧院建成。
哈尔滨工业大学主楼建成。

1966

《应县木塔》出版。
南宁体育馆建成。
中国人民解放军基建工程兵正式成立。
《建筑学报》复刊后再次停刊。

沿长江大桥远观桥头堡旧影

南京长江大桥桥头堡
Bridgeheads of Nanjing Yangtze River Bridge

1967

河南省体育馆建成。
援建项目几内亚人民宫建成。

南京长江大桥是长江上第一座由中国自行设计和建造的双层铁路、公路两用桥梁，在中国桥梁史乃至世界桥梁史上具有重要意义。1960年1月，大桥工程局委托中国建筑学会发动建筑设计单位和院校为桥头建筑征集设计方案。南京工学院（今东南大学）钟训正提出的复堡式红旗方案被采纳。“三面红旗”成为了那个时代的符号和标志。

As the first double-deck railway and highway bridge built independently of China, the Nanjing Yangtze River Bridge is of great significance in Chinese and the world's bridge history. In the same month, the Bridge Engineering Bureau entrusted the Architectural Society of China with soliciting the design proposal of the bridgehead from architecture design institutions and universities. The double-fort red-flag design of Zhong Xunzheng from Nanjing Institute of Technology (Southeast University now) was adopted. The "Three Red Flags" has become a symbol and mark of the time.

南立面旧影 / 东立面改造后局部 / 南立面外观

首都体育馆

Capital Indoor Stadium

首都体育馆是一座大型多功能体育馆，平面呈矩形，可同时进行24场乒乓球比赛，还可以进行冰球、体操等多项体育比赛以及集会和文艺演出。体育馆有观众席18000个，活动座席1200个。建筑屋顶为112.2米×99米的平板型双向空间网架，整体刚度好，用钢量少。该体育馆中建成了国内第一个室内冰球场，于1968年9月底完成。

The construction of the Capital Indoor Stadium was completed in September 1968. As a large rectangular multi-functional stadium, twenty-four of table tennis games can be held simultaneously, plus ice hockey, gymnastics, assembly and theatrical performance. It has a capacity of 18,000 seats and 1,200 moveable seats. Its roof, 112.2 meters long and 99 meters wide, is a flattop two-way space truss with excellent rigidity and steel effectiveness. The indoor ice hockey rink is the first in China.

1968

南京长江大桥建成通车。
刘敦桢（1897—1968）先生逝世。

1969

天安门城楼重建工程开工。
国家建设委员会、建筑工程部、建筑材料工业部的军管会联合向中央提出合并国家建委、建工部、建材部，成立国家基本建设委员会的报告。
浙江体育馆建成。

1970

根据中央发出的精简机构下放企业的文件，建工部、建材部与国家建委合并，原建筑工程部直属建筑施工、勘察设计、科学研究、大专院校等企事业单位，绝大部分下放地方领导。

1974

北京饭店
Beijing Hotel

北京饭店坐落于北京市东长安街，始建于1900年，是一家历史悠久的大型豪华饭店。饭店由东、中、西三幢楼组成，占地约4.2公顷。饭店中楼于1917—1919年由法国建筑师设计，建筑面积约15000平方米。饭店西楼建于1954年，建筑面积26000平方米。饭店东楼建于1974年，建筑面积约89000平方米。饭店的三幢建于不同年代的建筑有着各自不同的表现。

Erected on Dongchang'an Street, Beijing, built in 1900, Beijing Hotel is a luxury hotel with a long history. Covering an area of about 4.2 hectares, the hotel has a complex of three towers, the East, Central and West. Designed and constructed by a French architect during 1917-1919, floor area is 15,000 square meters. The West Tower was built in 1954, with a floor area of 26,000 square meters. The East Tower, covering a floor area of 89,000 square meters, was completed in 1974. Although the three towers were built in different times and in different styles.

北京饭店全景／北京饭店三期南立面

1971

国家建委在北京召开全国设计革命会议，重点批判基本建设方面的“大、洋、全”，提出进一步贯彻毛泽东主席关于开展群众性设计革命运动的指示，搞好设计战线的“斗、批、改”。

杭州机场候机楼建成。

郑州二七纪念塔建成。

1972

广州矿泉别墅建成。

台北国父纪念馆建成。

梁思成（1901—1972）先生逝世。

北京国际俱乐部建成。

1973

扬州鉴真纪念堂建成（1963年设计）。

全国几十家设计院联合编撰策划《建筑设计资料集》（第一版）。

芬兰建筑图片展举行。

1974

北京饭店东楼落成。

建筑科学研究院在北京召开全国住宅设计经验交流会。

北京建外外交公寓建成。

上海延安饭店建成。

南宁剧院建成。

1975

上海体育馆建成。

南京五台山体育馆建成。

国家建委召开了设计标准规范管理工作座谈会。

新疆驻京办事处建成。

北京动物园两栖爬行馆建成。

天津友谊宾馆建成。

1976

7月28日唐山发生大地震，百年工业城市顷刻间夷为平地。

北京长途电话大楼建成。

广州市白云宾馆建成。

北京协和医院门诊楼建成。

陕西户县农民画展览馆建成。

广州市矿泉旅舍建成。

1977

毛主席纪念堂落成。

合肥骆岗机场建成通航，这是我国第一座大型国际备降机场。

长沙火车站建成。

1978

全国科学大会在北京召开。建筑科技方面获奖项目有176项。

国家建委召开城市住宅工作会议。

中国建筑学会在南宁召开“文化大革命”后第一次建筑创作学术交流会议。

《建筑设计资料集》（第一版）正式出版。

杭州剧院建成。

团中央办公楼建成。

外贸谈判大楼建成。

1979

国务院设立国家建筑工程总局、国家城市建设总局。

国家建委发布《关于印发〈对全国勘察设计单位进行登记和颁发证书的暂行办法〉的通知》。这是中华人民共和国成立后的第一个勘察设计市场准入制度。

国家计委、国家建委、财政部发出《关于勘察设计单位实行企业化取费试点的通知》。根据通知，全国18家勘察设计单位成为全国首批企业化管理改革试点单位，由核拨事业费改为停拨事业费，收取设计费，采取自收自支、自负盈亏、自我约束、自我发展的企业化管理的经营模式。这是中华人民共和国成立后第一次实行设计收费制度。

刘敦桢遗著《苏州古典园林》出版。

我国第一幢整体预应力装配式板柱结构试验楼建成。

首都国际机场候机楼及配套工程竣工。

中国工程建设标准化委员会第一次全国代表大会在武昌召开。

1980

国家建工总局颁发直属勘察设计单位试行企业化收费暂行实施办法。

颁布《城市规划编制审批暂行办法》《城市规划定额指标的暂行办法》。

国家建工总局颁发《优秀建筑设计奖励条例》（试行），要求建工系统逐级推荐优秀设计，并规定以后每两年评选一次，评选范围为两年内投产的项目。

刘敦桢主编的《中国古代建筑史》出版。

中美合资建造的中国第一家中外合资饭店——北京建国饭店开工建设。

全国中、小型剧场方案设计竞赛评选会议在成都召开。

上海电信大楼建成。

西安钟楼饭店建成。

从广场远观纪念堂 / 纪念堂前雕像

1977

毛主席纪念堂

Chairman Mao Memorial Hall

毛主席纪念堂位于天安门广场，占地 5.72 公顷，其主体建筑为柱廊型正方体，南北正面镶嵌着镌刻有“毛主席纪念堂”六个金色大字的汉白玉匾额，44 根方形花岗岩石柱环抱外廊，雄伟挺拔，庄严肃穆，具有独特的民族风格。纪念堂打破了坐北朝南的传统，主立面朝北，与纪念碑为同一朝向。其平面布局严整对称，参观路线通畅。

Chairman Mao Memorial Hall, sits on Tian'anmen Square and covers an area of 5.72 hectares. The main structure is a colonnaded cubic construction. On both south and north sides there is a white-marble horizontal board with a gilded six-character inscription reading Chairman Mao Memorial Hall in Chinese. With 44 square granite columns supporting the portico, the entire structure looks magnificent, solemn and stately, with distinctive national characteristics.The memorial hall breaks the tradition of facing south. Instead, it faces north, in the same direction as the memorial monument.

改革繁荣创作

Booming Creation by Reform

1980—1999

改革开放，学习国外的先进思想和技术，不仅可以打开建筑界视野，也为建筑创作带来了新气象，出现了有当代意义的中国现代建筑。空前繁荣的时代，造就出一大批如：白天鹅宾馆、香山饭店、中国国际展览中心、陕西历史博物馆、国家奥林匹克中心、东方明珠、金茂大厦等新建筑。与此同时，吴良镛、齐康、张锦秋、程泰宁、何镜堂、马国馨、柴裴义、唐玉恩等中国建筑师的出现，也为中国现代建筑的地域创作提供了多方面、多色彩的先锋“示例”。虽面对出现的盲目模仿、良莠不齐的现象，但在20世纪建筑先驱们艰苦卓绝努力下的中国建筑总体的现代化前进脚步是值得敬畏的。

Since the reform and opening-up, learning foreign advanced thoughts and technologies has helped Chinese architects not only widen their horizon, but also brought new life into architectural creation, making modern Chinese architecture with contemporary reference be flourishing as never before and bringing about a large number of new buildings including White Swan Hotel, Fragrant Hill Hotel, China International Exhibition Center, Shaanxi History Museum, National Olympic Sports Center, Oriental Pearl Tower and Jinmao Tower. Meanwhile, such Chinese architects as Wu Liangyong, Qi Kang, Zhang Jinqiu, Chen Taining, He Jingtang, Ma Guoxin, Chai Peiyi and Tang Yu'en have provided multi-colored pioneering examples for regional creation of Chinese modern architecture in many aspects. In spite of the presence of blind imitation and uneven quality, it is awe-inspiring that the overall modernization of Chinese architecture moved ahead with the arduous efforts of architectural pioneers in the 20th century.

建筑与环境

松江方塔园

Songjiang Square Pagoda Park

方塔园位于上海市松江区，1981年初步建成。该园占地172.73亩，建筑设计采用以方塔为中心，塔殿不在同一轴线上等自由的布局方法，灵活的空间构成，表现出了宋文化的典雅、朴素。方塔园还有明代大型砖雕照壁、宋代望仙桥、明代兰瑞堂（又名楠木厅）、清代天妃宫等建筑。

The Pagoda Park is in Songjiang District of Shanghai and fundamentally completed in 1981. The park covers an area of 172.73 mu (1 mu= 667 m^2). Since the new architectural complex was centered on the pagoda, the layout did not use the central axis of pagoda-hall configuration, but a more flexible spatial composition to express the grace and simplicity of the Song's culture. A large brick carving screen wall of Ming was erected, in addition to Wangxian Bridge of Song Dynasty (Bridge of Gazing at Immortals Afar), Lanrui Hall of Ming Dynasty (also called Nanmu Hall), Tianfei Palace of Qing Dynasty (Palace of Heavenly Concubine), etc.

1981

国家建委印发了《对职工住宅设计标准的几项补充规定》。

国家建委在北京召开全国优秀设计总结表彰会议，会议评选出70年代国家优秀设计项目121个。

上海龙柏饭店建成。

《营造法式大木作研究》出版。

《全国工程建设标准设计管理办法》颁布。

《建筑工程技术干部职称业务标准的通知》发布。

全国农村住宅设计竞赛揭晓。

浙江大学图书馆建成。

南立面与湖面倒影

香山饭店
Fragrant Hill Hotel

香山饭店坐落于北京市香山公园内，设计者重视建筑与环境的和谐，如采用中国庭院式的布局，将“厌直不厌曲”的理念，“借景”“框景”及“隔”的处理手法以及对“空灵”感的追求诠释、展现得恰到妙处，并使江南民居简朴、雅致的轮廓和色调，北方四合院的形式和功能性与西方现代建筑交融在一起。

Fragrant Hill Hotel is located in Fragrant Hills Park, Beijing City. The hotel shows emphasis upon the harmony between architecture and environment. It adopts a layout typical of Chinese courtyards and epitomizes the concept of "preferring curvedness to straightness", the techniques of "borrowing views", "framing views" and "partitioning" and the pursuit of the impression of "spaciousness and airiness" (kong-ling). It successfully combines the simple, elegant contour and tone characteristic of folk architecture in Southeast China, the form and functionality of the quadrangles in North China, and some features of Western modern architecture.

1982

国务院设立城乡建设环境保护部。

国务院颁布《中华人民共和国文物保护法》。

中国建筑工程总公司成立。

国务院批转国家建委、国家文物局、国家城建总局联合所做的《关于保护我国历史文化名城的请示》，并公布了有重大历史价值和革命意义的24个城市为中国第一批历史文化名城。

“中国传统建筑图片展览”在香港展出。

著名建筑师杨廷宝（1901—1982）先生逝世。

北京图书馆东楼建成。

白天鹅宾馆

White Swan Hotel

白天鹅宾馆位于广州市沙面南侧，紧邻珠江白鹅潭。宾馆的公共空间如门厅、休息厅、咖啡厅、餐厅等临江设置，便于客人欣赏江景。主楼平面为“腰鼓”形，南北方向的阳台均由斜板构成，在阳光下产生明暗面，显得雅致精巧；外墙采用白色饰面，颇有白天鹅羽翼重叠之意，使建筑与环境融为一体。

White Swan Hotel is located in the south of Shamian Island, Guangzhou, adjacent to Bai'etan in the Zhujiang River. In the waterfront public facilities such as the lobby, lounge, café and restaurant, guests can enjoy a beautiful outlook of the river. The main structure has a plan resembling a waist drum. The two balconies in the south and north are made up of slanting planes so that in sunshine they have dark and bright sides and look elegant and ingenuous. The exterior walls are decorated with white on surface, which makes the building look like a white swan with folded wings and thus fit well with the environment.

1983

国家计委等部门联合发出关于勘察设计单位试行技术经济责任制的通知，将国家按人头多少拨给事业费，改为向建设单位收取勘察设计费。

童寯（1900—1983）先生逝世。

全国农村集镇剧场设计方案竞赛举办。

建设部颁发《建筑设计人员职业道德守则》。

成立全国通用建筑标准设计协作委员会。

瑞士文化基金会筹备的瑞士“1970—1980建筑图片展览会”在北京开幕。

《新建筑》在武汉创刊。

梁思成遗著《营造法式注释》出版。

南京金陵饭店建成营业。

首都规划建设委员会成立并举行第一次会议。

厦门高崎机场建成。

外立面

内庭院及廊道 / 建筑入口外景

武夷山庄
Wuyi Mountain Villa

武夷山庄坐落在著名的武夷山风景名胜区，地处闽江源头崇阳溪畔。其整体设计以“宜低不宜高，宜散不宜聚，宜土不宜洋”为原则，采用整体规划、分期实施、逐步调整的步骤。行列布置、群体的垂直线对组合建筑群有重要的意义，并尽可能地留出发展余地，是“新乡土主义”的经典代表作。

The Wuyi Mountain Villa is situated in the well-known resort area of the Wuyi Mountain, abutting the Chongyang Creek, the headwaters of the Minjiang River. The overall design was based on the principle of "rather be low than high, be more spacious than dense, and be more local than foreign". The entire planning was integral, phased, and stage-modified. The arrangement was by rows and columns and a collective vertical line, which is significant for composite architecture. Considerations for future extension were included. It is a classic of "Neo-Ruralism".

1984

国务院颁布《城市规划条例》。
《时代建筑》在上海创刊。
义县奉国寺大殿修缮工程动工。
在云南大理召开“中国传统民居建筑幻灯汇映会”。
现代中国建筑创作小组成立。
建设部副部长戴念慈，就国家允许开办个体建筑设计事务所问题，对《经济日报》记者发表谈话。
国家计委、城乡建设环境保护部发出了《关于印发〈工程承包公司暂行办法〉的通知》。
中法住宅学术讨论会在北京举行。
中美房屋建筑与城市规划技术讨论会在北京举行。

外景鸟瞰 / 入口空间组织

阙里宾舍

Queli Hotel

阙里宾舍西临孔庙，北临孔府，其设计以“甘当配角”的思想为指导，化整为零，严格限制高度，以与古建筑的体量、尺度相一致。房屋不超过两层，总轮廓线错落有致，采用民居形式，借鉴传统院落的组合形式，采用青筒瓦坡顶，灰砖及白粉墙，灰色斩假石梁、柱。

With the Temple of Confucius on the west and the Kong Family Mansion on the north, Queli Hotel, designed according to the principle of "performing a supporting role", has the same dimensions as ancient buildings. The houses are either one- or two-storied. The contour line rises and falls in an elegant way. Adopting the form of a traditional Chinese residential compound, it features a sloped roof paved with green pantiles, grey bricks, white walls, grey artificial-stone beams and pillars.

1985

我国第一部记载当代建筑业发展历程与建筑成就的大型工具书《中国建筑业年鉴》出版。

深圳南海酒店建成。

曲阜阙里宾舍建成。

“大地”建筑事务所成立，这是北京第一家中外合作经营的建筑设计单位。

中国建筑学会与建设部设计局在北京共同召开“繁荣建筑创作座谈会”。

首都规划建设委员会全体会议通过了《北京市区建筑高度方案》。

中国建筑历史研究座谈会在北京举行。

中国第一座伊斯兰文化中心工程在宁夏银川举行奠基典礼。

“电脑在建筑设计中的应用”学术交流会举行。

“繁荣建筑创作学术座谈会”在广州举行。这是自1959年上海“住宅建筑标准及建筑艺术问题座谈会”以后，第一次研究建筑创作问题的全国性专题会议。

中国国际展览中心 2~5 号馆

China International Exhibition Center Hall 2~5

中国国际展览中心主体为四个一字排开的边长为 63 米的大盒子，突显出其大空间、大尺度、大柱网的现代化特征。建筑师在每个方盒子的连接部位安排出入口，包含门厅、楼梯和走廊，出入口有凸出的拱形门廊，上方有圆弧形额枋，建筑外墙上部设有外凸的高窗。

The main body of China International Exhibition Centre is an array of four 63-meter long boxes with the modern feature of large space, dimensions, and hypostyle. The architect puts the entrances/exits, including the lobby, staircase and corridor at the connecting points between boxes. The entrances/exits have an arched vestibule with an arched architrave. Embossed windows are out of the exterior walls.

入口设计 / 内部展览布局 / 入口侧外观

悼念广场中的标志碑 / "倒下的 300000 人" 雕塑 / 入口侧立面

侵华日军南京大屠杀遇难同胞纪念馆（一期）

Memorial Hall of the Victims in Nanjing Massacre by Japanese Invaders (I)

侵华日军南京大屠杀遇难同胞纪念馆坐落在南京江东门街，该馆的所在地是侵华日军南京大屠杀江东门集体屠杀遗址和遇难者丛葬地。纪念馆占地 3 万平方米，采用灰白色大理石垒砌而成，气势恢宏、庄严肃穆，是一处以史料、文物、建筑、雕塑、影视等综合手法全面展示南京大屠杀特大惨案的专史陈列馆。其 2007 年进行了扩建。

The Memorial hall of the victims in Nanjing Massacre by Japanese invaders, or Nanjing Massacre Memorial Hall, lies on Jiangdongmen Street, Nanjing. It is also the site of Jiangdongmen Massacre and the burying ground of the killed. Covering an area of 30,000 square meters, it is built with greyish white marble. It is a splendid, solemn and stately, exhibition hall comprehensively exhibiting the Nanjing Massacre by historic materials, cultural relics, architecture, sculptures, audios and videos.The memorial was extended in 2007.

大堂内景 / 外立面远观

新疆人民会堂

Xinjiang People's Hall

新疆人民会堂坐落于乌鲁木齐市友好北路，由主体和副体组成，主体包括有3160个座位的观众大厅及舞台配套设施，副体有多功能厅，13个地、州、市会议厅以及餐厅、厨房等。主体前厅设有自动扶梯和大楼梯，还有题为“天山之春”的大理石壁画。建筑立面方圆组合，高低错落有致，极富地方特色。经过改造后，原有风格已有改变。

With the main mass that has a 3,160-seat auditorium and stage settings and amenities that include a multi-functional hall, 13 regional, provincial, and municipal meeting halls, as well as restaurant and kitchen, the Xinjiang People's Hall stands on Youhaobei Road in Urumqi. A large staircase supports an escalator for the traffic of the front lobby of the main hall, where a marble fresco themed the "Spring in Mountain Tianshan" is eye-catching. The architectural facade is a combined sphere and cube in a corbel form, revealing distinctive regional characteristics. After the transformation, the original style has been changed.

1986

国务院批准第一批全国烈士纪念建筑物保护单位。民政部公布32个保护单位名单。
全国首届建筑教育思想讨论会在南京召开。
深圳国际贸易中心大厦建成。
首次国家级优秀建筑设计、优质工程评选活动举行。
《中外合作设计工程项目暂行规定》发布。
唐山抗震纪念碑建成。
自贡恐龙博物馆建成。
杭州黄龙饭店建成。
中国银行大楼建成。

1987

北京图书馆新馆建成。
上海铁路新客站建成。
《住宅建筑设计规范》颁布实施。
全国勘察设计工作会议和中国勘察设计协会第一届理事会议在北京召开。
中国建筑业联合会设立建筑工程鲁班奖。
我国第一个现代化彩色电视制作播出中心——中央彩色电视中心竣工。
国家标准《中小学校建筑设计规划》颁布实行。
天津水晶宫饭店建成。
华亭宾馆建成。
深圳科学馆建成。
北京国际饭店建成。
云谷山庄建成。

主立面及广场布置

雨花台烈士陵园

Yuhuatai Memorial Park of Revolutionary Martyrs

雨花台烈士陵园是1949年后中国建设规模最大的纪念性陵园。陵园以主峰为中心形成南北向中轴线，通过建筑与自然的围合、建筑的围合、半人工围合，直到开敞的空间，渐次达到空间序列的高潮。建筑采用带有传统特色的现代建筑形式，对传统建筑形式加以变化，以简洁的手法表达传统建筑精神。

Yuhuatai Memorial Park of Revolutionary Martyrs is the largest memorial park in China built after 1949. The memorial park is laid out along the north-south axis with the main peak at the center. Initially there are enclosures formed by buildings and nature, then enclosures by buildings, semi-artificial enclosures and lastly open spaces. In this way, the complex culminates in terms of spatial arrangement. The form of traditional architectural elements is modified to convey the spirit of traditional architecture with simple techniques.

1988

中共中央办公厅国务院办公厅联合通知，严格控制建立纪念设施。

清华大学建筑学院成立。

《中国大百科全书　建筑·园林·城市规划》出版。

海峡两岸建筑专家、学者首次在香港聚会。

北京市评选出北京80年代十大建筑，北京图书馆新馆等10项建筑荣获北京80年代十大建筑称号。

深圳华联大厦建成。

1989

中国第一部《无障碍设计规范》由建设部颁布实施。

全国第一本由设计院创办的建筑设计类杂志《建筑创作》创刊。

西藏布达拉宫维修保护工程启动。

天津国际展览中心建成。

新疆吐鲁番新宾馆建成。

四川省体育馆建成。

新华社业务技术楼建成。

大连银帆宾馆建成。

西安古都大酒店建成。

上海市建设工程“白玉兰奖”创办。

全景鸟瞰

国家奥林匹克体育中心

National Olympic Sports Center

▌ 国家奥林匹克体育中心位于北京市朝阳区安定路一号，是为 1990 年在北京召开的第十一届亚运会而建设的大型体育中心。总建筑面积 12 万平方米，包括体育场、体育馆、游泳馆、曲棍球场及练习场、医务测试中心及练习场地等设施。其在总体规划和单体建筑设计中着重考虑了功能与形式的结合、现代与传统的结合、环境与建筑的结合。

▌ The National Olympic Sports Center, located at No. 1 Anding Road, Chaoyang District, Beijing City, is a large sports center built for the 1990 Beijing Asian Games, with a total floor area of 120,000 square meters, it consists of the stadium, gymnasium, natatorium, hockey field, medical testing center and training ground. The overall plan and individual structure design show consideration for combination of functionality and form, modernity and tradition, environment and architecture.

1990

首批全国工程勘察设计大师名单公布（120人）。建筑界20位设计大师是：齐康、孙芳垂、孙国城、严星华、杨先健、佘浚南、陈植、陈浩荣、陈登鳌、陈民三、张镈、张开济、张锦秋、赵冬日、徐尚志、容柏生、黄耀莘、龚德顺、熊明、戴念慈。

《中华人民共和国城市规划法》公布。

在北京举办“国际体育建筑学术交流会”。

上海市评出“上海十佳建筑”和“上海30个建筑精品”。曲阳新村、上海体育馆、上海游泳馆、上海展览馆、淀山湖大观园、华亭宾馆、静安希尔顿酒店、铁路上海站、华东电业大楼、延安东路隧道荣获“上海十佳建筑”称号。闵行一条街等荣获“上海建筑精品”称号。

新锦江大酒店建成。

第十一届亚运会在北京举行，一大批新建体育设施投入使用。

桂林桂湖饭店建成。

陕西历史博物馆
Shaanxi History Museum

陕西历史博物馆位于陕西省西安市大雁塔西北侧，馆区占地65000平方米，文物库区面积为8000平方米，展厅面积为11000平方米。博物馆由馆名碑池、主馆、库区、东南角楼、西南角楼、临时陈列厅、行政用房、业务用房等仿唐风格的建筑组成。

Shaanxi History Museum is located to the northwest of the Big Wild Geese Pagoda (Dayan Pagoda) in Xi'an of Shaanxi Province. The museum covers an area of 65,000 square meters, of which 8,000 square meters are warehouses of cultural relics. The exhibition hall is of a total area of 11,000 square meters. The museum is a Tang Dynasty-style building complex consisting of the pool of the tablet inscribed with the name of the museum, main hall, the warehousing area, southeast corner tower, southwest corner tower, temporary exhibition hall, office building, business rooms, etc.

全景鸟瞰 / 入口大厅

1991

陕西历史博物馆建成。
首届国际城市建设技术交流和展览会在北京举行。
潘天寿纪念馆建成。
华夏艺术中心建成。
北京炎黄艺术馆建成。
天津国际大厦建成。
南开大学东方艺术楼建成。
广东国际大厦建成。
西安喜来登大酒店建成。

1992

《成立中外合营工程设计机构审批管理的规定》颁布。
建设部确定全国32个大型勘察设计单位参加现代企业制度试点。
中国国际工程咨询协会成立。
建设部授予17个单位“全国工程勘察设计先进单位”称号。
深圳发展中心大厦建成。
北京市人民检察院办公楼建成。
西安城堡大酒店建成。

1993

建设部颁发《私营设计事务所试点办法》，并在上海、广州、深圳三地实施设立私营建筑设计事务所试点。
第一次建筑与文学研讨会在南昌召开。
中宣部办公楼建成。
北京建材经贸大厦建成。
北京林业大学主楼建成。
浙江省博物馆建成。

东侧鸟瞰 / 设计模型

北京菊儿胡同新四合院

Beijing Ju'er Hutong New Courtyard House

北新四合院位于北京市旧城区，属于旧城改造项目。设计师提出了“类四合院”式的新街坊体系，用高低错落的住宅、过街楼等围合成新的四合院，体现了建筑的“有机更新”。菊儿胡同是北京旧城改造的一次成功实验。

Located in the old town of Beijing City, the Ju'er Hutong New Courtyard House is one of the old-town transformation programs. The designer proposes the new neighborhood system featuring quasi-siheyuan. Residences and street-crossing buildings of varying heights enclose new siheyuans. It shows the “organic update” of architecture. The Ju'er Hutong New Courtyard House is a successful experiment of the transformation of the old town of Beijing.

侧立面外观

西汉南越王墓博物馆

Museum of the Nanyue King Mausoleum in Western Han Dynasty

博物馆占地14000平方米，整体布局以古墓为中心，上盖覆斗形钢架玻璃防护棚，象征汉代帝王陵墓覆斗形封土。墓的东边为三层的综合陈列楼，北边为两层的主体陈列楼，环绕的回廊上下沟通，将三座建筑物连成一个整体。

The museum covers an area of 14,000 square meters. Its overall plan is centered on the ancient mausoleum that is covered with a bucket-shape steel glass protective casing to imitate the same shape of the grave mound used to cover the ancient emperor's graves in Han Dynasty. On the east side of the mausoleum is a three-storied complex exhibition building, and the north side is the main exhibition building. A corridor rings around connect the three buildings as a whole.

1994

张镈著《我的建筑创作道路》出版。

《关于工程设计单位改为企业若干问题的意见》发布。

全国建筑师管理委员会成立。

《关于建立注册建筑师制度及有关工作的通知》发布。

上海大剧院竣工。

“'94首都建筑设计汇报展”在北京举行。

甲午海战馆建成。

北京丰泽园饭店建成。

天津体育馆建成。

东方明珠广播电视塔

Oriental Pearl Tower

远观东方明珠广播电视塔

东方明珠广播电视塔位于上海浦东新区陆家嘴，塔高约 468 米，1995 年 5 月投入使用。建筑整体似火箭冲天。全塔 11 个高低错落的球体串联起蓝天、绿地，中间的两个大球体宛如红宝石，晶莹夺目，电视塔的球体与近旁上海国际会议中心的两个地球球体形成“大珠小珠落玉盘”的意境。电视塔还建有观光层、旋转餐厅、陈列馆等设施。

The 468-meter high Oriental Pearl (Radio & TV) Tower is located at Lujiazui in Pudong New Area. It was put into use in May 1995. The overall architecture is like a thrusting rocket pointing to the sky. The tower has 11 spheres, high or low, connecting the blue sky with the green meadow. Two big spheres in the middle are like two rubies glittering brilliantly, which together with the two earth spheroids of the nearby Shanghai International Convention Center to create an atmosphere resembling “Pearls, big or small, drops on jade plate”. The TV tower has a sightseeing platform, revolving restaurant, and an exhibition hall.

1995

《国家安居工程实施方案》发布。

上海市评选出20世纪90年代十大新景观，浦江双桥等十项工程光荣当选。

国务院颁布《中华人民共和国注册建筑师条例》。

第一次一级注册建筑师考试在全国31个考场举行，9100人参加考试。

1996

《中华人民共和国注册建筑师条例实施细则》发布，并于1996年10月1日施行。

北京城市公厕建设文化展览在中国革命博物馆开幕。

国际建筑师协会金奖评出6个奖项。清华大学教授吴良镛获建筑评论奖和建筑教育奖。这是中国建筑师第一次获此大奖。

《住宅产业现代化试点技术发展要点》（试行）颁布。

金茂大厦与周边建筑组团

金茂大厦
Jin Mao Tower

金茂大厦位于上海市浦东新区陆家嘴金融贸易区，高 420.5 米，1999 年建成，地上 88 层，若加上尖塔的楼层共有 93 层，地下 3 层，有多达 130 部电梯与 555 间客房。大厦由塔楼、裙房、地下室三部分组成，裙房为水平向建筑，与塔楼形成了强烈对比。曲线型屋顶与向上内收的外墙为简洁的外形增加了立体感。

In Shanghai Pudong New Area abutting Lujiazui Finance and Trade Zone by the Bund, the Jin Mao Tower is 420.5 meters high. It was completed in 1999. The tower is of 88 floors above ground, or 93 if adding the spire, and three below the ground level, which require 130 elevators and 555 guest rooms to serve its needs. The tower consists of the tower body, annex and basement. The annex is a group of horizontal architecture in strong contrast with the tower. The curved and tapered upward exterior walls add a stereoscopic impression to its otherwise simple and monotone appearance.

1997

《中国生态住宅技术评估手册》出版发行。

东南大学建筑系举行建系70周年庆祝活动。

1998

《中华人民共和国建筑法》开始实施。

首届全国电脑建筑画大赛获奖作品展在中国美术馆开幕。

上海现代建筑设计（集团）有限公司正式挂牌。

国家大剧院方案竞赛在北京中国革命历史博物馆举行公开展览。

第一届中国建筑史学国际研讨会在北京举行。

建设部发布《中小型勘察设计咨询单位深化改革指导意见》。

1999

第20届世界建筑师大会在北京举行，两院院士吴良镛代表国际建协发表《北京宣言》。

中华人民共和国50年上海经典建筑评选揭晓，金茂大厦等10项建筑获金奖，上海商城等10项建筑获银奖，华亭宾馆等10项建筑获铜奖。

建筑师

Architects

项目评选
历程
实录

陈嘉庚

入选项目：
集美学村（构想、创办）

陈嘉庚（1874—1961），著名爱国华侨、企业家、教育家、慈善家、社会活动家，福建省泉州府同安县集美社人（今厦门市集美区）。

1913年，陈嘉庚在家乡泉州府同安县集美社创办集美学村，1921年创办了有文、理、法、商、教育等5院17系的厦门大学。这是当时全国唯一一所独资创办的大学。陈嘉庚不是建筑师，但其以远见卓识规划建设了中国东南沿海的两处堪称经典的校园建筑组群。陈嘉庚近代校园规划设计理念的形成有美国建筑师墨菲带来的影响及帮助。

墨　菲

入选项目：
清华大学早期建筑
未名湖燕园建筑

亨利·墨菲（Henry Killam Murphy，1877—1954），美国建筑师，1918年7月在上海外滩开办了个人事务所，20世纪上半叶在中国规划设计了雅礼大学（1914年）、清华大学（1916年）、金陵女子大学和燕京大学（1921—1926年）等多所重要大学的校园，并在南京等地设计了一批建筑作品，如紫金山国民革命军阵亡将士公墓纪念塔、南京国民政府铁道部大楼等，并主持了南京的城市规划，是当时中国建筑古典复兴思潮的代表性人物。

邬达克

入选项目：
国际饭店
上海外滩建筑群（部分）

邬达克（Laszlo Hudec，1893—1958），出生于奥匈帝国兹沃伦州（Zolyom）首府拜斯泰采巴尼亚（Besztecebanya）的一个建筑世家，21岁毕业于匈牙利皇家约瑟夫理工大学（今布达佩斯理工大学）建筑系。后在上海拥有了自己的建筑设计事务所。邬达克寄居上海长达30年，他先后设计了60余项建筑作品，国际饭店、大光明电影院等近1/3的建筑被列为上海优秀历史建筑。

吕彦直

入选项目：
中山陵
中山纪念堂

吕彦直（1894—1929），安徽滁县（今滁州市）人，1894年出生于天津。作为中国近代杰出的建筑师，吕彦直在短暂的一生中弘扬民族文化，在中国近代建筑史上写下了辉煌的一页。由他设计、监造的南京中山陵和主持设计的广州市中山纪念堂都是富有中华民族特色的大型建筑组群，是我国近代建筑中融汇东西方建筑技术与艺术的代表作。他是“中国固有式建筑”流派的奠基人之一，在建筑界产生了深远的影响。鉴于他对建造孙中山陵墓的杰出贡献，在他逝世后，南京国民政府曾明令全国予以褒奖，并在陵园立碑纪念。

茅以升

入选项目：
钱塘江大桥

茅以升（1896—1989），字唐臣，江苏镇江人。土木工程学家、桥梁专家、工程教育家，中国科学院院士、美国工程院院士、中央研究院院士。1916年毕业于西南交通大学（时称交通部唐山工业专门学校），1919年获美国卡耐基理工学院博士学位。回国后历任交通大学唐山工学院教授、院长，钱塘江大桥工程处处长，铁道科学研究院院长等职。曾主持修建钱塘江大桥（中国铁路桥梁史上的一座里程碑），参与设计武汉长江大桥，尤以钱塘江大桥最富传奇色彩。

梁思成

入选项目：
人民英雄纪念碑
鉴真纪念堂

梁思成（1901—1972），杰出的中国建筑历史学家、建筑教育家、建筑师。1927年6月毕业于美国宾夕法尼亚大学建筑系，获硕士学位。1928年回国后创办东北大学建筑系并任系主任。1930年加入朱启钤创办的中国营造学社，1946年创办清华大学建筑系并任教授、系主任。1955年当选中国科学院技术科学部委员。留有《中国建筑史》《清式营造则例》等巨著。梁思成的主要设计作品有人民英雄纪念碑、北京大学地质馆、北京大学女生宿舍、鉴真纪念堂（莫宗江补充设计）。梁思成1949年之后在设计、规划上的成果的学术意义在于：在实践中检验中国古代建筑原则与文化精神应用于现代中国建设的实际效果。

杨廷宝

入选项目：
北京和平宾馆
紫金山天文台
北京市百货大楼
中央大学

杨廷宝（1901—1982），生于河南南阳，1925年毕业于美国宾夕法尼亚大学建筑系，后在基泰工程司工作，1949年后任南京大学、南京工学院建筑系教授、系主任。1979年起兼任南京工学院建筑研究所所长，同年任江苏省副省长。1957年和1961年两次当选为国际建筑师协会副主席。1955年当选为中国科学院技术科学部委员。杨廷宝先生是业内在设计造诣上久负盛名的大师级人物，更是用设计作品探索中国古典建筑、民间建筑与西方科技最新理念相结合的大家。杨廷宝先生设计完成的项目有100多项。

陈　植

入选项目：
上海外滩建筑群（部分）

陈植（1902—2002），生于浙江杭州，1928年获美国宾夕法尼亚大学建筑系建筑硕士学位。1929年回国后到东北大学建筑系执教，1931—1952年同建筑师赵深、童寯在上海合办华盖建筑师事务所。中华人民共和国成立后，陈植历任华东建筑设计公司总建筑师等职，先后参与了上海中苏友好大厦工程的建设，设计了鲁迅墓及纪念馆，主持了闵行一条街、张庙一条街等重点工程设计，为上海的城市建设与发展做出了贡献。他主持和指导的苏丹友谊厅设计赢得了良好的国际声誉。

刘开渠

入选项目：
人民英雄纪念碑

刘开渠（1904—1993），江苏徐州府萧县人（今安徽），著名雕塑家。早年毕业于北平美术学校，后赴法国，归国后历任杭州艺术专科学校（中国美术学院）教授、中国美术馆馆长、中国美术家协会副主席。其早年曾创作淞沪战役阵亡将士纪念碑等一批反映抗战题材的艺术作品。后与梁思成、莫宗江等组成人民英雄纪念碑浮雕创作小组，具体负责纪念碑基座浮雕的总体构思，并创作其中的胜利渡长江解放全中国、支援前线、欢迎解放军等浮雕。以其名字命名的“刘开渠奖”为中国雕塑界的最高奖项。

张　镈

入选项目：
人民大会堂
民族文化宫
北京友谊宾馆
北京自然博物馆
北京饭店（东楼）

张镈（1911—1999），山东省无棣县人。1934年毕业于中央大学建筑系。后在基泰工程司从事建筑设计工作。1951年3月从香港回京任北京市建筑设计院总建筑师。1990年被评为首批全国工程勘察设计大师。完成百余项工程设计，1949年后的代表作有人民大会堂、民族文化宫、新侨饭店、友谊医院、文化部办公楼、积水潭医院、光明日报社、钓鱼台国宾馆18号楼等。1994年出版《我的建筑创作道路》系以个人传记及口述体讲述建筑创作生涯的第一书。

黎伦杰

入选项目：
重庆人民解放纪念碑

黎伦杰(1912—2001)，广东番禺人，1937年毕业于广东省立勷勤大学建筑工程学系。1939年5月，黎伦杰担任中山大学建筑学系助教，1940年3月前往重庆。在重庆期间，黎伦杰先后担任中国新建筑社事务所技师，《新建筑》杂志主编，重庆大学建筑系讲师、副教授等职，并受聘担任重庆都市计划委员会工程师，为重庆抗战胜利纪功碑的设计者。其在教学、实践的同时，发表了一系列研究专著和论述。黎伦杰战后回到广州，与郑祖良合组新建筑工程司开展设计业务。

华揽洪

入选项目：
北京儿童医院

华揽洪（1912—2012），生于北京，1928年赴法留学，获法国国授建筑师称号（DPLG）。1951年回国后任北京市都市计划委员会总建筑师，1954年任北京市建筑设计研究院总建筑师，1977年退休后移居法国，1981年任中国建筑学会名誉理事。曾任中国建筑学会第七至九届海外名誉理事。其代表作品有北京儿童医院，北京市社会路住宅楼，北京市幸福村住宅小区，位于北京市右安门、复兴门、西直门、三里屯等地的多处住宅及办公建筑，巴黎Glaciere街中国留学生招待所，中国驻法国巴黎领事馆改造，中国驻联合国机构代表住宅楼改造，中国驻法国大使馆文化处等。

徐　中

入选项目：
天津大学主楼

徐中（1912—1985），江苏常州人。1935年毕业于中央大学建筑系，获学士学位。1937年毕业于美国伊利诺伊大学，获建筑设计硕士学位。1940—1949年任中央大学建筑系教授，1954年被任命为天津大学建筑系主任，为天津大学建筑系的发展呕心沥血，倾注了满腔热情，为教育事业贡献了毕生精力。曾设计南京国立中央音乐学院校舍、南京馥园新村住宅、南京交通银行行长钱新之住宅、北京商业部进出口公司办公楼、对外贸易部办公楼、天津大学主楼等。

杨作材

入选项目：
延安革命遗址

杨作材（1912—1989），江西德化（今九江）人。1936年毕业于武汉大学。1938年入延安抗大学习。曾任延安自然科学院校务处处长、中央军委办公厅副主任、冀热辽军区政治部秘书处处长、中共热中地委敌工部部长、东北林务总局局长。1949年后，历任重工业部办公厅副主任，冶金工业部设计司司长，国家建委副主任，国家计委副主任、顾问，是第五届全国政协委员。在延安期间，曾独立设计并指挥施工建造了杨家岭中央大礼堂、中央办公厅办公楼、王家坪中央军委机关礼堂。

张开济

入选项目：
中国革命历史博物馆
天安门观礼台
钓鱼台国宾馆
北京天文馆及改建工程

张开济（1912—2006），浙江杭州人。1935年毕业于中央大学建筑系。先后在上海公和洋行设计部等就职。1949年年底任北京市建筑设计院总建筑师，曾任中国建筑学会副理事长。1990年被评为首批全国工程勘察设计大师。代表作品有中国革命历史博物馆、天安门观礼台、钓鱼台国宾馆、北京“四部一会”建筑群、中央民族学院和北京天文馆等。2000年获中国首届梁思成建筑奖。晚年在《北京晚报》开辟专栏，连载建筑批评类文章，后结集出版。

洪　青

入选项目：
西安人民大厦
西安人民剧院

洪青（1913—1979），早年留学欧洲，毕业后曾在上海美术专科学校担任美术系教授。1950年举家迁至陕西西安，加入当时西北建筑公司（中建西北设计院前身）从事设计工作。几十年中他为西安的建筑事业做出卓越贡献，作品享誉海内外。仅中华人民共和国成立初期（1950—1959年）洪青在西安就设计了西安人民大厦、西安人民剧院、陕建集团办公楼、陕西宾馆、西安新华书店钟楼店旧址、西安交大主群楼、西安邮政局大楼、陕西纺织供销公司办公楼、西北一印旧址等建筑。

张家德

入选项目：
重庆市人民大礼堂

张家德（1913—1982），生于四川省威远县。1934年毕业于中央大学建筑工程系，1935—1938年就职于军政部营造司、南京炮兵学校工程处，先后设计了中央大戏院、沙利文舞楼、聚兴城银行等工程。1941—1949年在重庆创办家德建筑事务所。重庆人民大礼堂是他最杰出的代表作。中华人民共和国成立后，在北京市建筑设计院研究所、信息所工作，参与了部分重点项目的设计与研究。据相关媒体和网站的报道，他曾在城市建设部建筑设计院工作，之后任中国建筑科学研究院副总工程师。

莫伯治

入选项目：
白天鹅宾馆
泮溪酒家
广州白云山庄
西汉南越王墓博物馆

莫伯治（1914—2003），生于广东东莞。1936年毕业于广州中山大学工程院土木建筑系，曾任广州市规划局总建筑师、广州市规划局技术总顾问等，1955年创立莫伯治建筑师事务所。1994年被评为全国工程勘察设计大师，1995年当选中国工程院院士，2000年获首届梁思成建筑奖。代表作品有广州泮溪酒家、广州白云山庄、白云山双溪别墅、广州矿泉别墅、白云宾馆、白天鹅宾馆、岭南画派纪念馆、西汉南越王墓博物馆等。

赵冬日

入选项目：
人民大会堂

赵冬日（1914—2005），生于辽宁彰武。1941年毕业于日本东京早稻田大学建筑系。中华人民共和国成立后，历任北京市规划局总建筑师，北京市建筑设计院总建筑师。1990年被评为首批全国工程勘察设计大师。从事建筑教育、城市规划、建筑设计数十年，积累了丰富的实践经验，对北京市城市规划和建筑设计做出了突出贡献。代表作品有中国伊斯兰教经学院、全国政协礼堂、中直礼堂、人民大会堂（方案）、天安门广场规划、北京同仁医院。2000年获得中国首届梁思成建筑奖。

冯纪忠

入选项目：
松江方塔园

冯纪忠(1915—2009)，生于河南开封。1934—1936年就读于圣约翰大学土木系，后赴欧洲留学，1948年8月起任同济大学教授，1951—1953年创办上海群安建筑师事务所。1955年12月—1983年任同济大学建筑系系主任。1956年创立中国第一个城市规划专业，1958年开设城市规划专业园林方向。他先后参加南京及上海的都市规划，设计了武汉东湖客舍、武汉医院（现同济医学院附属医院）主楼等在业内产生了重大影响的建筑。20世纪70年代末冯纪忠规划设计了松江方塔园。

徐尚志

入选项目：
成都锦江宾馆

徐尚志（1915—2007），生于四川成都。1939年毕业于重庆大学土木工程系，20世纪40年代初与戴念慈先生等组建重庆怡信工程司。中华人民共和国成立后，任重庆建筑公司工程师、西南建筑设计院建筑师、中国建筑西南设计院总建筑师。1990年被授予首批全国工程勘察设计大师称号。曾主持设计重庆市劳动人民文化宫礼堂、重庆体育馆、重庆宾馆、重庆剧场、成都锦江宾馆、锦江大礼堂等数十项工程建筑，在现代化技术与传统风格相结合等方面进行过颇有成就的探讨和实践。

陈登鳌

入选项目：
北京火车站

陈登鳌（1916—1999），江苏无锡人，1937年毕业于上海沪江大学城中区商学院建筑系。1937年至1948年在上海及南京从事建筑设计工作。1949年至1982年在新乡、北京从事建筑设计和技术管理工作，历任建设部建筑设计院副总建筑师、建设部建筑设计院顾问总建筑师等。1990年被评为全国工程勘察设计大师。主持设计的主要建筑作品有：中央军委景山机关宿舍大楼、洛阳三厂住宅区第一期工程、北京火车站(与南京工学院合作)、几内亚共和国国家大会堂。

林乐义

入选项目：
北京电报大楼

林乐义（1916—1988），生于福建南平，1937年毕业于上海沪江大学，抗战胜利后到美国佐治亚理工学院研究建筑学，1950年回国，任建筑工程部北京工业建筑设计院总建筑师，建设部建筑设计院总建筑师等职。主要代表作有桂林艺术馆、广西大学、南京储汇大楼、广西忠烈祠、北京首都剧场、北京电报大楼、北京国际饭店、中国驻波兰大使馆、青岛一号工程、北京东郊使馆区工程、郑州“二七”烈士纪念塔、中南海怀仁堂和紫光阁改建工程。他主持编写的《建筑设计资料集》（第一版）已成为建筑和其他专业设计人员常备的重要工具书。

莫宗江

入选项目：
人民英雄纪念碑
鉴真纪念堂

莫宗江（1916—1999），广东新会人。1931年加入中国营造学社，作为建筑学宗师梁思成先生的主要助手，考察了山西、河北、河南、陕西、北京等地的大量古代建筑，1943年完成梁思成所著《中国建筑史》的插图绘制工作，后随梁思成创办了清华大学建筑系，任教授。中华人民共和国成立后，梁思成、林徽因、莫宗江、高庄等密切合作，共同完成了清华大学建筑系的国徽设计方案。1951年与梁思成、刘开渠等合作设计人民英雄纪念碑，并在梁思成先生逝世后补充完善了梁思成于1963年所作的鉴真纪念堂初步设计。

贝聿铭

入选项目：
香山饭店

贝聿铭(1917—　)，祖籍苏州，先后在麻省理工学院和哈佛大学攻读建筑学，获美国建筑学会金奖、法国建筑学会金奖、日本帝赏奖和普利兹克奖。贝聿铭先生承担设计建造的有约翰·肯尼迪图书馆、北京香山饭店、法国卢浮宫改扩建工程、苏州博物馆等建筑工程，被誉为“现代主义建筑的最后大师”。但从香山饭店、苏州博物馆等作品看，他由现代主义的“反传统”要素过渡为“借鉴传统、兼收并蓄”，自有其“特立独行于各个流派之外”的独步天下之处。

巫敬桓

入选项目：
北京市百货大楼
北京和平宾馆

巫敬桓（1919—1977），生于重庆。1945年毕业于中央大学建筑系，留校任杨廷宝教授的助教，1951年，与夫人张琦云随杨廷宝教授到北京，加盟兴业公司建筑设计部。后随设计部并入北京市建筑设计院。参与设计了北京和平宾馆，主持设计了北京市百货大楼、八面槽百货商场、人民日报社建筑群、中国文联办公楼。此外，还参与设计了新侨饭店，主持设计了汽车局建筑、全国工商联办公楼、石油学院教学楼、建外中小型使馆、阜外南营房住宅楼、国务院宿舍楼、河南医学院建筑、北京师范学院楼群等。

戴念慈

入选项目：
中国美术馆
阙里宾舍
北京展览馆
北京饭店（西楼）

戴念慈（1920—1991），生于江苏省无锡市。1942年毕业于中央大学建筑系。1944—1948年于重庆、上海的兴业建筑师事务所及信诚建筑师事务所就职。1950年后于北京中直修建办事处、建筑工程部设计院、中国建筑科学研究院工作，任总建筑师。1982—1986年任城乡建设环境保护部副部长。1983—1991年任中国建筑学会理事长。1991年11月当选中国工程院院士。主要代表作品有中国美术馆、北京饭店西楼、斯里兰卡国际会议大厦、阙里宾舍等重要工程。他的贡献在于以其建筑设计实践去探索建筑学和建筑师应该如何为现代中国建设服务。

吴良镛

入选项目：
北京菊儿胡同新四合院

吴良镛（1922— ），生于江苏，1944年毕业于重庆中央大学建筑系，1950年获美国匡溪艺术学院硕士学位，1946年起协助梁思成教授创建清华大学建筑系，曾任清华大学建筑系副主任、主任，国际建筑师协会副主席，世界人居学会主席。中国科学院和中国工程院两院院士。1999年在第二十届世界建筑师大会上，他代表国际建协宣讲了面向21世纪的《北京宣言》。20世纪90年代初，他设计建成的北京菊儿胡同新四合院，先后获国家和建设部的优秀设计奖，亚洲建协建筑设计金牌奖和联合国世界人居奖。荣获2011年国家最高科学技术奖等重要奖项。

龚德顺

入选项目：
建设部办公楼

龚德顺（1923—2007），生于北京，籍贯浙江杭州。1945年毕业于天津工商学院建筑工程系，毕业后从事建筑设计工作。曾任国家建筑工程总局副总建筑师、城乡建设环境保护部设计局局长、中国建筑学会秘书长、中国建筑学会建筑师分会会长等职。1990年被授予全国工程勘察设计大师称号。代表作品有建设部办公楼、军委政治大学、蒙古人民共和国百货大楼、深圳华夏艺术中心等民用建筑以及洛阳轴承厂、拖拉机装配车间、蒙古人民共和国毛纺厂、人民日报彩色印刷厂等工业建筑。

张德沛

入选项目：
北京友谊宾馆

张德沛（1925—2015），祖籍山东，抗战时曾为国民革命军伞兵部队翻译。1946年9月—1950年夏在清华大学营建系学习，为第一届毕业生 。1953年开始在北京市建筑设计研究院工作。主要设计作品有玉泉山后山总体设计附属建筑（包括道桥）、北京工业学院附中教学楼、北京友谊宾馆（建筑专业负责人）、北京市市委党校主体工程、印度大使官邸、东城区结核病防治所、东城区工人俱乐部、北京市密云水库管理办公楼、三里屯使馆区（多座使馆的设计人）等。

关肇邺

入选项目：
清华大学图书馆
北京大学图书馆

关肇邺（1929— ），祖籍广东南海。1952年毕业于清华大学建筑系，毕业后留校任教至今。1995年当选为中国工程院院士。2000年被授予全国工程勘察设计大师称号，并获得首届梁思成建筑奖。现任清华大学建筑学院教授、博士生导师。长期致力于文化、教育建筑的设计和研究，主持完成了清华大学图书馆、北京大学图书馆的扩建工程等设计。

钟训正

入选项目：
南京长江大桥桥头堡

钟训正（1929— ），生于湖南武冈。1952年毕业于南京大学建筑系，曾先后任教于湖南大学、武汉大学，1954年始任教于南京工学院（现东南大学）建筑系至今。1997年11月当选为中国工程院院士。钟训正院士20世纪50年代末参加了北京火车站综合方案的设计，1968年完成的南京长江大桥桥头堡建筑方案位居全国竞赛第一，后经周恩来总理选定为实施方案。他还主持设计了无锡太湖饭店新楼、甘肃画院及海南三亚金陵度假村等建筑。

齐　康

入选项目：
雨花台烈士陵园
侵华日军南京大屠杀遇难同胞纪念馆

齐康（1931— ），生于江苏。1952年毕业于南京工学院建筑系（现东南大学建筑学院），1952年毕业后留校任教，历任教研组主任、系主任、系党总支书记，南京工学院副院长，建筑研究所副所长、所长，建筑设计研究院总顾问等职。1965年当选中国建筑学会理事、常务理事，之后各届连任理事至今。1990年被选为中国勘察设计大师（建筑），1993年被选为中国科学院院士， 2000年获得首届梁思成建筑奖。主持设计了许多工程项目。

张锦秋

入选项目：
陕西历史博物馆

张锦秋(1936—)，生于四川成都。1954—1960年就读于清华大学建筑系，1961—1964年为清华大学建筑系建筑历史和理论研究生，师从梁思成、莫宗江教授。1966年至今在中国建筑西北设计研究院从事建筑设计，担任院总建筑师。其间，主持设计了许多有影响的工程项目，如：西安大雁塔景区的三唐工程、陕西历史博物馆和西安群贤庄小区、西安钟鼓楼广场、陕西省图书馆和美术馆群体建筑、黄帝陵祭祀大殿、大唐芙蓉园等。代表性著作有《长安意匠——张锦秋建筑作品集》。1990年被授予首批全国工程勘察设计大师称号，1994年被遴选为中国工程院首批院士， 2000年获首届梁思成建筑奖。

何镜堂

入选项目：
西汉南越王墓博物馆

何镜堂（1938— ），生于广东，1961年获华南工学院（现华南理工大学）建筑学学士学位，1965年获华南工学院（现华南理工大学）建筑学硕士学位，建筑学家，中国工程院院士。现任华南理工大学建筑设计研究院院长、总建筑师，华南理工大学教授、博士生导师。设计了2010年上海世博会中国馆、侵华日军南京大屠杀遇难同胞纪念馆扩建工程等一大批有重大影响的精品工程。1994年获全国工程勘察设计大师称号，1999年当选为中国工程院院士。获首届梁思成建筑奖。

柴裴义

入选项目：
中国国际展览中心2~5号馆

柴裴义（1942— ），生于天津，1967年毕业于清华大学土木建筑系建筑专业，1974年至今在北京市建筑设计研究院有限公司任总建筑师，现任顾问总建筑师。1981—1983年赴日本东京丹下健三建筑设计研究所研修。2004年获第四届全国工程勘察设计大师称号，2008年获第五届梁思成建筑奖。他设计的主要作品有中国国际展览中心2~5号馆、孟加拉国际会议中心、国际投资大厦、加蓬国民议会大厦、雾凇宾馆等。

马国馨

入选项目：
国家奥林匹克体育中心

马国馨（1942— ），生于山东省济南市。1959—1965年就读于清华大学建筑系，后到北京市建筑设计研究院工作至今。现为院总建筑师。1981—1983年在日本东京丹下健三都市建筑研究所研修。1991年获清华大学工学博士学位。1994年被授予全国工程勘察设计大师称号。1997年当选为中国工程院院士。长期在设计第一线从事设计工作，先后参与和负责了国家和北京市的多项重点工程并多次获奖。出版《丹下健三》《日本建筑论稿》《体育建筑论稿——从亚运到奥运》等专著。

唐玉恩

入选项目：
上海外滩建筑群（部分建筑改建）

唐玉恩（1944— ），生于重庆，1967年毕业于清华大学建筑学专业，1978年考取同济大学建筑学专业研究生，1981年获硕士学位，后至上海建筑设计研究院工作，1999年起任院总建筑师。2004年被授予全国工程勘察设计大师称号，2012年获得第六届梁思成建筑奖提名奖。主要作品有上海图书馆新馆、上海现代建筑设计大厦、复旦大学逸夫楼、三亚金茂希尔顿大酒店、上海浦东新区少年宫与图书馆、上海罗店美兰湖国际会议中心、上海柏树大厦、上海虹桥新区新世纪大厦、连云港市财政干部培训中心、江苏石油化工学院图书馆等。

庄惟敏

入选项目：
中国美术馆（改建）

庄惟敏（1962— ），生于上海，1985年从清华大学建筑学专业本科毕业， 1992年3月从清华大学博士毕业。现任清华大学建筑设计研究院院长、总建筑师，清华大学建筑学院院长、教授、博士生导师。2008年获全国工程勘察设计大师称号，著有《建筑策划导论》等，曾主持中国美术馆改建工程，世界大学生运动会游泳跳水馆，2008北京奥运会射击馆、飞碟靶场和柔道跆拳道馆等重大工程的设计工作。

赵元超

入选项目：
西安人民大厦（改扩建）

赵元超(1963—)，生于西安，1985年毕业于重庆建工学院建筑学专业，1988年从重庆建工学院建筑设计及其理论方向研究生毕业，获建筑学硕士学位。研究生毕业后进入中国建筑西北设计研究院工作，现任中国建筑西北设计研究院总建筑师。主要设计有西安行政中心、西安城墙南门广场、西安浐灞商务行政中心、西安人民大厦整体改扩建、延安大剧院、延安市历史博物馆、延安市民文化中心和行政中心、宁夏党委办公区、西安火车站改扩建工程等。

附录：首批中国20世纪建筑遗产名录（按所获票数为序）

序号	名称
1	人民大会堂
2	民族文化宫
3	人民英雄纪念碑
4	中国美术馆
5	中山陵
6	重庆市人民大礼堂
7	北京火车站
8	清华大学早期建筑
9	天津劝业场大楼
10	上海外滩建筑群
11	中山纪念堂
12	北京展览馆
13	中央大学旧址
14	北京饭店
15	国际饭店
16	中国革命历史博物馆
17	天津五大道近代建筑群
18	集美学村
19	厦门大学早期建筑
20	北京协和医学院及附属医院
21	武汉国民政府旧址
22	孙中山临时大总统府及南京国民政府建筑遗存
23	清华大学图书馆
24	北京友谊宾馆
25	武汉大学早期建筑
26	鉴真纪念堂
27	武昌起义军政府旧址
28	香山饭店
29	国立紫金山天文台旧址
30	未名湖燕园建筑群
31	汉口近代建筑群
32	北京和平宾馆
33	白天鹅宾馆
34	毛主席纪念堂
35	徐家汇天主堂
36	北京大学红楼
37	长春第一汽车制造厂早期建筑
38	北京电报大楼
39	圣·索菲亚教堂
40	北京“四部一会”办公楼
41	上海展览中心
42	雨花台烈士陵园
43	黄花岗七十二烈士墓园
44	阙里宾舍
45	钱塘江大桥
46	重庆人民解放纪念碑
47	西泠印社
48	金陵大学旧址
49	松江方塔园
50	钓鱼台国宾馆
51	侵华日军南京大屠杀遇难同胞纪念馆（一期）
52	首都剧场
53	武汉长江大桥
54	北京天文馆及改建工程
55	陕西历史博物馆
56	国家奥林匹克体育中心
57	北京市百货大楼
58	北京工人体育场
59	南岳忠烈祠
60	延安革命旧址
61	江汉关大楼
62	上海鲁迅纪念馆
63	广州白云山庄
64	东方明珠上海广播电视塔
65	天安门观礼台
66	北洋大学堂旧址
67	建设部办公楼
68	天津大学主楼
69	北京菊儿胡同新四合院
70	北京儿童医院
71	武夷山庄
72	中国共产党第一次全国代表大会会址
73	西汉南越王墓博物馆
74	佘山天文台
75	国民政府行政院旧址
76	同济大学文远楼
77	曹杨新村
78	首都体育馆
79	金茂大厦
80	泮溪酒家
81	中国营造学社旧址
82	南京长江大桥桥头堡
83	重庆黄山抗战旧址群
84	国民参政会旧址
85	清华大学1~4号宿舍楼
86	西安人民大厦
87	北京自然博物馆
88	华新水泥厂旧址
89	中国国际展览中心2~5号馆
90	西安人民剧院
91	北京大学图书馆
92	同盟国中国战区统帅部参谋长官邸旧址
93	重庆抗战兵器工业旧址群
94	南泉抗战旧址群
95	马可·波罗广场建筑群
96	新疆人民会堂
97	南京西路建筑群
98	成都锦江宾馆

《文化传播·创意设计·展览系列》丛书编委会

学术顾问　吴良镛　谢辰生　关肇邺　李道增　傅熹年　彭一刚　陈志华　张锦秋　程泰宁
何镜堂　郑时龄　费　麟　刘景樑　王小东　王瑞珠　黄星元

名誉主编　单霁翔　马国馨　修　龙

丛书主编　金　磊

丛书编委（按姓氏笔画排序）　王建国　王时伟　付清远　孙宗列　孙兆杰　朱　颖　伍　江　刘伯英　刘克成
刘若梅　刘　谞　庄惟敏　邵韦平　邱　跃　何智亚　张立方　张　宇　张　兵
张　杰　张　松　张爱林　李秉奇　李　沉　杨　瑛　陈　薇　陈　雳　陈　雄
季也清　赵元超　徐　锋　郭卫兵　殷力欣　周　岚　周　恺　孟建民　金卫钧
常　青　崔　愷　梅洪元　奚江琳　路　红　韩振平

《致敬中国建筑经典——中国20世纪建筑遗产的事件·作品·人物·思想》编委会

主　　编　金　磊

执行主编　朱有恒

执行编辑　殷力欣　李　沉　彭长昕　苗　淼　崔　勇　董晨曦　郭　颖　林　娜　金维忻

图片提供　中国建筑学会建筑摄影专业委员会等

“致敬中国建筑经典：中国20世纪建筑遗产的事件 · 作品 · 人物 · 思想”特展

顾　　问　单霁翔 马国馨 修龙

策 展 人　金磊

执行策展人　李沉　苗淼　朱有恒　董晨曦　金维忻　刘晓乐

图书在版编目（CIP）数据

致敬中国建筑经典：中国 20 世纪建筑遗产的事件·作品·人物·思想 / 中国文物学会 20 世纪建筑遗产委员会，《中国建筑文化遗产》编辑部主编. -- 天津：天津大学出版社，2018.5
（文化传播·创意设计·展览系列）
ISBN 978-7-5618-6119-6

Ⅰ. ①致… Ⅱ. ①中… ②中… Ⅲ. ①建筑—文化遗产—中国—20 世纪 Ⅳ. ① TU-092

中国版本图书馆 CIP 数据核字（2018）第 083243 号

Zhijing Zhongguo Jianzhu Jingdian: Zhongguo 20 Shiji Jianzhu Yichan de Shijian Zuopin Renwu Sixiang

策划编辑：韩振平　郭　颖
责任编辑：刘　浩
装帧设计：朱有恒

出版发行　天津大学出版社
地　　址　天津市卫津路92号天津大学内（邮编：300072）
电　　话　发行部：022-27403647
网　　址　publish.tju.edu.cn
印　　刷　北京华联印刷有限公司
经　　销　全国各地新华书店
开　　本　165mm×229mm
印　　张　8.5
字　　数　100千
版　　次　2018年5月第1版
印　　次　2018年5月第1次
定　　价　48.00元